AF386643

LYMPHOCYTE STIMULATION

Differential Sensitivity to Radiation

Biochemical and Immunological Processes

LYMPHOCYTE STIMULATION

Differential Sensitivity to Radiation

Biochemical and Immunological Processes

Edited by

AMLETO CASTELLANI

CNEN
Rome, Italy

PLENUM PRESS · NEW YORK AND LONDON

Library of Congress Cataloging in Publication Data

Main entry under title:

Lymphocyte stimulation.

"Proceedings of the European Molecular Biology Organization lecture course...
held September 11−19, 1978, in Frascati, Rome, Italy."
Includes index.
1. Lymphocyte transformation–Congresses. 2. Deoxyribonucleic acid synthesis–
Congresses. 3. Deoxyribonucleic acid repair–Congresses. I. Castellani, Amleto.
II. European Molecular Biology Organization.
QR185.8.L9L87   599.01'13.   80-19883
ISBN 978-1-4684-6999-8

ISBN 978-1-4684-6999-8   ISBN 978-1-4684-6997-4 (eBook)
DOI 10.1007/978-1-4684-6997-4

Collection of papers presented at the European Molecular Biology Organization
lecture course on Lymphocyte Stimulation, held in Frascati, Rome, Italy,
September 11−19, 1978.

PREFACE

    This book is a collection of some of the papers presented at
the EMBO Lecture Course on "Lymphocyte stimulation: differential
sensitivity to radiation; biochemical and immunological properties."

    The Course was organized with the aim of fostering interactions
between photoradiobiologists and immunologists interested in the
problem of DNA damage and repair studied at the lymphocyte level.
The papers presented in this book are mainly centered on the problem
of radiation sensitivity of lymphocytes in relation to DNA repair
phenomena.

    The radiation biology of human lymphocytes is dominated by two
phenomena:

    (a) high radiosensitivity of lymphocytes which die in interphase
    (b) PHA-induced relative radioresistance of those cells which,
        after stimulation, escape the interphase death and eventually
        die in mitosis.

    These phenomena constitute a good system to study some of the
factors which control the response of human cells to irradiation.
In addition it is possible to correlate the development of the
relative radioresistance in PHA-stimulated lymphocytes with the
biochemical changes connected with the transformation processes.

    The papers presented in this book constitute a real contribution
to the scientific knowledge in this field of research and suggest
that lymphocytes could be a very interesting test material useful
for measuring the DNA repair capability of human cells to furnish
an indication of individual radiosensitivity in man.

                         A. Castellani
                         Division of Radiation Protection
                         CNEN - CSN Casaccia, Rome, Italy

CONTENTS

LYMPHOCYTE STIMULATION BY NONSPECIFIC MITOGENS

Bilha Schechter

Department of Cell Biology, The Weizmann Institute of
Science, Rehovot, Israel

INTRODUCTION

Studies on the mitogenic activation of lymphocytes began in
1960 with the discovery by Nowell that a lectin extracted from the
plant Phaseolus vulgaris (phytohemagglutinin, or PHA) transformed
small resting lymphocytes into proliferating lymphoblasts (1).
Since then, many other lectins (2), as well as agents other than
lectins, have been found to stimulate lymphocytes non-specifically.
The binding of stimulating agents, i.e., mitogens to the surface of
lymphoid cells induces a sequence of surface and metabolic events
which culminate in DNA synthesis and mitosis. It is generally
accepted that the gross morphological and biological changes
following mitogenic stimulation are similar to those which follow
antigen-induced immune reactions. However, mitogenic stimulation
"by-passes" the requirement for specific recognition that characterizes immunological activation. Mitogens were, therefore, found to
be useful in studies on the mechanism of lymphocyte activation and
in the evaluation of the biological and immune competence of lymphocytes. Assays for mitogen-induced lymphocyte transformation
provide a very useful means of assessing immunological competence
in various types of immunodeficiency states, both in experimental
and in clinical work (3, 4).

A large number of agents are known to be mitogenic (Table 1).
These include lectins from plant and animal sources, bacterial
products, antibodies to lymphocyte surface components, calcium
ionophores, oxidizing agents and various enzymes. Some of the
mitogens are known to be able to distinguish between the two major
subpopulations of lymphocytes, i.e., the bone marrow - (B) and the

Table 1.   <u>Lymphocyte Mitogens</u>

1. <u>Lectins</u> (see Table 3)

2. <u>Bacterial Products</u>

Lipopolysaccaride (Lipid A)	B cells
Staphylococcal enterotoxin B	B cells
Tuberculin, Purified Protein Derivative (PPD)	B cells
Streptolysin S	B cells

3. <u>Antibody Reagents</u>

Anti-immunoglobulin sera	B cells
Anti-lymphocytic sera	
Carbohydrate-specific antibodies	
anti-α_2-macroglobulin	
anti-β_2-microglobulin	B cells

4. <u>Miscellaneous</u>

Sodium metaperiodate $NaIO_4$	T cells
Phorbol esters (APA) (stimulate cGMP synthesis)	T cells
Ca^{++}-ionophore A-23187 (transports Ca^{++} into cells)	
Metal ions (zinc, mercury, nickel)	
Polyene antibiotics	B cells
Sulfated polyanions (e.g., dextran sulfate)	B cells
Proteolytic enzymes (trypsin, papain)	
Galactose oxidase (after neuraminidase treatment)	T cells
Antigens	T cells
Allogeneic cells (MLC)	T cells

thymus (T)-derived cells. The interaction between lymphocytes and
a mitogenic agent results in the initiation of a chain of events
(Table 2), the earliest of which are changes in membrane fluidity
and permeability, redistribution of surface receptors, and accelerated turnover of membrane components. These are followed by
intracellular events such as stimulation of histone acetylation,
phosphorylation of nuclear proteins and modification of lipids and
carbohydrate metabolism. Stimulation of RNA and protein synthesis
is detected at 4 to 6 h after contact with potent mitogens such as
PHA or Concanavalin A (Con A), and these are followed by active
DNA synthesis. The cells then undergo morphological changes, turn
into large blast-like cells and enter mitosis. In addition to these
changes which characterize most cells undergoing active growth,
lymphocyte stimulation is accompanied by production and release of
biologically active products typical for immune-activated lymphocytes. Such products are the lymphokines, which play a role in
cell-mediated immune responses (5-7) and are secreted by activated

Table 2. Events Following Lymphocyte Stimulation

<u>Within 60 minutes</u>

Clustering (patching) of surface receptors
Increased membrane permeability and enhanced uptake of
  ions (Ca^{++}, K^{+})
Accelerated turnover of membrane phospholipids
Changes in the cAMP/cGMP ratios

<u>A few hours</u>

Stimulation of acetylation of histones
Phosphorylation of nuclear proteins
Modification of lipid and carbohydrate metabolism
Stimulation of protein synthesis
Stimulation of RNA synthesis

<u>48 hours</u>

Active DNA synthesis and mitosis
Morphological changes (blast formation)
release of biologically active products (lymphokines)

<u>72—96 hours</u>

Immunoglobulin production by B cells
Cytotoxicity of T cells

thymus-derived lymphocytes (T cells), and immunoglobulins which are
produced and/or secreted by bone marrow-derived lymphoid
cells after 72-96 h of incubation with B cell mitogens (8, 9).
Nonspecifically active cytotoxic T lymphocytes appear after 4-5 days
of incubation with T cell mitogens (10).

It is accepted that lymphocyte activation is initiated by the
binding of the mitogenic molecule through its binding site(s) to
receptors on the cell surface. However, the nature of these receptors and the relationship between their passive binding to the
activating ligand and the crucial steps in the induction of activation have not been elucidated. Most studies on the mechanism of
lymphocyte activation were done with plant lectins, since those are
better defined with respect to their binding specificity than other
mitogens. Lectins are characterized by their ability to bind
sugars and to agglutinate erythrocytes of different animal species.
The sugars with which they can interact are typical constituents of
glycoproteins and glycolipids present on cell surfaces (2), including those of lymphoid cells. Most lectins bind lymphoid cells.
Nonetheless, only some lectins induce lymphocyte activation, and
although they can bind to both B and T cells, mitogenic lectins

mostly stimulate T rather than B lymphocytes. One approach to
studying the mechanism of lymphocyte activation is to search for
properties of interaction with cell surface receptors that will
distinguish mitogenic from non-mitogenic interactions.

BINDING SPECIFICITY

Most lectins interact preferentially with a single sugar structure, e.g., D-galactose,N-acetyl-D-galactoseamine, D-mannose,D-
glucose, or N-acetyl-D-glucoseamine (Table 3). Binding specificity
cannot account for the differences between mitogenic and non-
mitogenic lectins since most specificity groups include both types
of lectins. This may imply that a large variety of receptors is
involved in transmitting the mitogenic signal. It is, however,
possible that only a limited number of subclasses of receptors on
the cell surface is responsible for mitogenic interactions. This
view may be supported by the notions that some lectins are known to
interact best with complex carbohydrate structures that occur in
glycoprotines and glycolipids, and that for some lectins the specificity is broader and includes a number of closely related sugars.
Therefore in certain cases the sugar with which a mitogenic lectin
interacts best is not necessarily the one through which it imparts
its mitogenic trigger.

Soybean agglutinin (SBA), for example, is specific to N-acetyl-
D-galactoseamine (11) but can also bind, although with a lower
affinity, D-galactose. SBA stimulates lymphocytes from pig but not
from mouse, rat or human. The latter three species, although able
to bind SBA, can be stimulated by the lectin only after neuraminidase
treatment of cells (12). Since such treatment removes sialic acid
residues and exposes galactose-like sites, it has been suggested
that the increase in mitogenic activity is due to the binding of SBA
to new receptors made available by the neuraminidase treatment. The
stimulation by SBA of neuraminidase-treated lymphocytes could be
due to an increase in the number of available receptors or more
likely to the exposure of receptors through which stimulation occurs.
Treatment with neuraminidase is also required for stimulation of
rat and human lymphocytes by peanut agglutinin (PNA) (13), a lectin
specific for the disaccaride Gal-β(1- β)-GalNAc, and for D-galactose
although with a lower binding affinity. It is possible that a sub-
class of receptors on the cell surface with the terminal sequence
sialic acid→galactose→ is important in mitogenic activation. This
possibility is supported by the results of Novogrodsky et al. (reviewed in Ref. 14), who showed that NaIO$_4$ oxidation of sialic acid
residues, or β-galactosidase oxidation of galactosyl residues exposed after neuraminidase treatment induced extensive blastogenesis.
Such treatments yielded aldehyde moieties which could interact with

Table 3. <u>Mitogenic and Non-Mitogenic Plant Lectins</u>*

Source of lectin	Sugar speci-ficity	Valency	Mitogenic activity
Phaseolus vulgaris (red kidney bean), PHA	Gal-NAc		+ T, CRT
Glycine max (soybean), SBA		2	-
(SBA)$_n$		4–n	+ T-mouse
Phaseolus linensis (lima bean) I		2	±
II		4	+
Dolichos biflorus (horse gram)		4	-
Wistaria floribunda haemaggl. (WFH)			-
Wistaria floribunda mitogen (WFM)			+
Helix pomatia (garden snail)		6	-
Archis hypogaea (peanut), PNA	Gal	2	+ T-rat, human, Nase
(PNA)$_n$		n	+ T-rat, human, mouse Nase
Ricinus communis (castor bean), RCA		2	-
Abrus precatorius (jequirity bean)		2	+
Bandeirea simplicifolia		4	-
Pseudomonas aeruginosa			-
Canavalia ensiformis (jack bean) Con A	Glc,Man	4	+ T, CRT + CST
Succ- Con A		2	+
Lens culinaris (lentil)		2	+ T
Pisum sativum (garden pea)		2	+
Vicia faba (broad bean)			+
Triticum vulgare (wheat), WGA	(Glc-NAc)$_2$	4	-
Solanum tuberosum (potato)		2	
Phytolacca americana (pokeweed), PWM Pa-1	Unknown	**	+ B, T
Pa-2 to Pa-5		†	+ T

*References for the various lectins are listed in Table I of Ref. 2.

**Molecular weight (22000)$_n$.

†Molecular weight 19,000–31,000.

<u>Abbreviations</u>: Gal-NAc, N-acetyl-D-galactoseamine; Gal, D-galactose; Glc, D-glucose; Man, D-mannose; Glc-NAc, N-acetyl-D-glucoseamine; T, thymus-derived lymphocytes; B, bone marrow-derived lymphocytes; CRT, cortisone-resistant thymocytes; CST, cortisone-sensitive thymocytes; Nase, neuraminidase-treated.

other functional groups on the cell surface, such as free amino, alcoholic, or thiol groups, and the cross-linked structures thus formed might play a role in the initiation of the process of activation. Removal by β-galactosidase of galactose residues from neuraminidase-treated cells abolished almost completely the response of the cells to SBA, PNA and β-galactosidase oxidation, but not to Con A. Glycoprotein receptors, such as those carrying the terminal sequence NANA-Gal-β(1→4)-GlcNAc-Man-Man- may serve as multipotential receptors for lymphocyte activation by $NaIO_4$ oxidation (of intact cells) or by SBA, PNA and β-galactosidase (after neuraminidase treatment).

Glycoproteins isolated from solubilized membrane preparations of lymphocytes by affinity chromatography with different lectins revealed a similar profile and multiple lectin binding specificities. These glycoproteins may be distinguished by different affinities for a certain lectin, and the receptor which mediates transformation may have a much higher affinity for the mitogen. It is still not clear whether different mitogenic lectins share a common receptor molecule, whether receptors for a certain mitogen are homogenous or whether T lymphocytes possess receptor molecules for T cell mitogens that are not present on B lymphocytes, and vice versa. Although it is intriguing to speculate that a common receptor molecule mediates activation of T lymphocytes by a variety of ligands (antigens, lectins or anti-lymphocyte antisera), the evidence and nature of such a receptor are still a matter of controversy. The difference in binding specificity between ligands that mediate transformation and those that fail to transform is also unclear. It must be assumed that mitogenic lectins differ from non-mitogenic lectins in their ability to interact with the appropriate carbohydrate structures which occur in glycoproteins or glycolipids on the cell surface. The wide range of specificities of different mitogens suggests that the induction of proliferation is not aboslutely dependent on the specificity of a certain cell surface receptor. Properties of interaction other than binding specificity must, therefore, underlie the triggering process by the cells.

BINDING AFFINITY AND RECEPTOR HETEROGENEITY

Both mitogenic and non-mitogenic lectins bind to lymphocytes, implying that binding alone is not necessarily followed by activation. Binding properties such as affinity to cell surface receptors or the total number of receptors available for binding could account for the differences between the two groups of lectins. Studies with radioactively labeled lectins showed, however, that most lymphocytes bind 10^6–10^7 lectin molecules per cell, regardless of the lectin or lymphoid cell population. Similar numbers of

lectin molecules were bound to B and T cells, using T cell mitogenic
lectins such as PHA, Con A and lentil lectin (15-18). Also, lectins
that are not mitogenic at all, such as Rincinius communis agglutinin
(RCA), wheat germ agglutinin (WGA) (18, 19), Wistaria floribunda
hemagglutinin (WFH) (20) and agglutinin from Agaricus bisporus (21).
bind to the surface of human or mouse lymphocytes to a similar
extent as mitogenic lectins such as PHA (18, 21, 22), Con A (18, 19,
23) and lentil lectin (18, 24, 25). Such values were obtained when
measured at lectin saturation. It seems, however, that the majority
of the binding sites present on the cell do not play a role in the
initiation of transformation, since binding of a lectin to only a
small number of receptors is necessary for stimulation. At concentrations of Con A, PHA, lentil lectin or WFM that gave maximal
stimulation of lymphocytes, only a small fraction (0.3—15%) of the
total number of receptors for these lectins were occupied (17, 20,
23, 26-28). It is not known whether the occupied receptors are
qualitatively different from the unoccupied ones, but it has been
suggested that the former have a higher affinity for the mitogen
than the latter (28). The question is, therefore, whether interaction of cells with lectins at low lectin concentrations may reveal
differences between mitogenic and non-mitogenic lectins. Studies
with radioactively labeled lectins demonstrated that the binding
constants (association constants or Ka) for most lectins range
between 10^6 and 10^7 whether or not they are mitogenic (reviewed in
Ref. 2). Differences between these two groups of lectins could,
however, be demonstrated by plotting the Scatchard plots (29) of
the binding data of lectins to lymphocytes (30). Non-mitogenic
lectins such as native (unaggregated) SBA and PNA exhibited linear
saturation patterns, indicating that each molecule interacts with a
single receptor only and that all receptor sites on the cell and all
binding sites on the lectin are homogeneous. Deviation from
linearity in the binding curves was observed with mitogenic lectins
such as PHA, (SBA)n and (PNA)n (n = polymerized and multivalent)
suggesting that due to receptor hetcrogeneity binding at low site
occupancy is accelerated. Another possibility is that mitogenic
lectins exhibit positive cooperativity, that is, the initial interaction between the lectin and cell surface receptors results in an
increased lectin—receptor association. If cooperative binding
indeed characterizes mitogenic interactions, then it does not seem
to be a function of the properties of the lectin, since no
cooperativity was observed in binding of lectins to saccharides in
solution (31, 32). It is rather a function of the lectin—cell
interaction. Thus, PNA is not mitogenic, neither does it exhibit
positive cooperativity in binding to mouse T splenocytes, whereas
mitogenic stimulation and cooperativity were observed with this
lectin upon binding to rat hydrocortisone-resistant thymocytes (30).
The range where the accelerated binding occurs falls below or
within the mitogenic concentration of the lectin. Positive cooperativity in binding of mitogenic molecules to cell surface receptors
can be explained either by an increase in the affinity of the

receptors to the lectin or by an increase in the number of available receptor sites. Both types of changes may be due to conformational changes in membrane components or to their redistribution in the membrane facilitated by the fluid nature of the latter. Correlation between increased membrane fluidity shortly after interaction with lectins, and subsequent mitogenic activity was indeed observed with respect to a number of lectins. An increase in membrane fluidity was observed in lymphocytes treated with PHA, Con A (33), WFM and lentil lectin but not when treated with the non-mitogenic WFH and Sophora japonica agglutinin (20). It is suggested that the accelerated binding of mitogenic lectins at low site occupancy may reflect either interactions with high affinity receptors or alterations in cell membrane structure and organization which may be essential events in the stimulation process.

LECTIN VALENCY

The number of available binding sites occupied by the lectin molecule seems to be an important determinant of the ability of the lectin to stimulate lymphoid cells. We found that the mitogenic activity of SBA is dependent on the presence of lectin aggregates formed in lectin preparations upon lyophilization and storage. After fractionation into unaggregated (divalent), dimeric (tetravalent) and polymeric (multivalent) fractions, it was found that the unaggregated lectin stimulated neitehr pig lymphocytes nor neuraminidase-treated mouse cells, while the tetravalent and multivalent fractions were mitogenic to both cell types (34). These results suggest that SBA must have at least four sugar binding sites in order to be able to stimulate lymphocytes.

A similar observation was made with the lectin from lima bean, which, similarly to SBA, is specific for N-acetyl-D-galatoseamine. Two hemagglutinating lectins were purified from lima bean, a tetramer with four saccharide binding sites and a dimer with two binding sites. The tetravalent species in this case was severalfold more potent as a mitogen than the divalent lectin (35). The requirement for multivalent interactions of receptor sites could also explain the stimulation of neuraminidase-treated mouse splenocytes and thymocytes by aggregated PNA, but not by unaggregated divalent PNA (36), the stimulation of dinitrophenyl-modified thymocytes with divalent, but not monovalent antidinitrophenyl antibodies (37) and the triggering of B cells by insolubilized T cell mitogens (38).

In the case of Con A, the conversion of the native tetravalent lectin (specific to mannose-like residues) to a divalent lectin did not affect its mitogenic potency (39), although the two lectin species differed in some other qualities. Con A exhibited a typical biphasic dose-response curve which peaked at 1—10 µg/ml and declined

rapidly with increasing doses of mitogen, whereas the dose response
curve of the divalent Con A did not decrease up to a concentration of
200 µg/ml. Also, the divalent Con A, in contrast to the tetravalent
lectin, did not induce cap formation, neither did it restrict the
mobility of the receptors at higher concentrations (see below).
Increasing the valency of divalent Con A by adding antibodies against
Con A to cells that have bound divalent Con A restored both of these
activities. It thus appears that in the case of Con A, divalency is
sufficient to achieve receptor clustering which seems to be necessary for induction of stimulation. An apparent exception to the
requirement for multivalent interactions is the finding that monovalent fragments of antibodies which react with cell surface carbo-
hydrates and a monovalent Con A derivative were mitogenic to mouse
spleen cells (40-42). These observations can still be explained
considering the role of macrophages in mitogenic stimulation (43-46).
It has been shown that highly purified macrophage-depleted lymphocytes did not respond to PHA and Con A. Activation by these lectins
could be obtained both when lymphocytes pre-treated with mitogen
were cultured in the presence of untreated macrophages and when
mitogen-pretreated macrophages were incubated with untreated lymphocytes (46). It is assumed that mitogenic molecules are adsorbed on
macrophages and may thus be "presented" to the lymphocytes in a
higher local concentration and in a multivalent form, even when the
mitogen itself is of low valency. The requirement for multivalent
interactions suggests that crosslinking and clustering of lectin
receptors on the lymphocyte surface is required for the generation
of the triggering signal.

RECEPTOR DISTRIBUTION AND MOBILITY

Due to the fluid mosaic of the membrane, surface antigens are
believed to be mobile (47) and immunoglobulins and other receptors
in the lymphocyte membrane may be cross-linked and redistributed by
specific divalent antibodies. Redistribution induced by such ligands
consists first of patching or clustering of receptors by diffusion
and cross linkage and then capping or accumulation of the patches at
one pole of the cell (48). Similar observations were made with
fluorescent lectin derivatives or with ferritin- or peroxidase-
conjugated lectins showing that lectin receptors can readily undergo
redistribution on the cell surface by lateral movement in the plan
of the membrane to form patches and caps (49, 50). The caps may
then fall off or be ingested by the cell. It could be assumed that
patching, capping or ingestion might be a crucial step in the
initiation or triggering of the chain of events leading to stimulation. It was, however, found that ingestion into the cell subsequent to binding is not essential for mitogenic stimulation since
immobilized PHA, PWM, Con A and lentil lectin could activate lymphocytes (15, 38, 51-53). Also, no correlation was found between cap-

ping and stimulation. Thus, Con A, PHA and lentil lectin formed
caps on both B and T cells, although only T cells were stimulated
(15, 56). Mitogenic concentrations of PWM did not induce cap
formation (55), whereas cap formation was induced by the non-
mitogenic lectin from the mushroom <u>Agaricus bisporus</u> (56).

The mobility of receptors for one lectin as evidenced by capping may be unrelated to that of receptors for another lectin on
the same cell (57). Thus, redistribution of Con A receptors was
not affected by the distribution of surface immunoglobulins by anti-
immunoglobulin and vice versa (55, 56). If, however, Con A was
added to mouse splenic lymphocytes at doses greater than those
needed for optimal stimulation, receptor mobility was restricted
and cap formation of Con A receptors and immunoglobulin was
inhibited. There seems to be a correlation between restriction of
receptor mobility by lectins and their mitogenic activity (59).
Such inhibition was observed with six mitogenic lectins including
PHA but not with several non-mitogenic lectins. These observations
suggest that the interaction of a mitogen with the appropriate cell
surface receptor is signalled through the membrane to the inside
cell membrane complex. Edelman et al. (60) have proposed the
hypothesis that binding of a mitogen such as Con A causes structural
alterations of a common protein anchorage system associated with
microtubules and microfilaments inside the cell. The mobility of
some other receptors is restricted, since they are anchored to such
structures. This hypothesis is supported by the finding that drugs
affecting microtubules and microfilament assembly, such as col-
chicine, colcemide, vinblastine and vincristine can partially reverse
the effect of Con A induced restricted mobility and permit the
formation of Con A and anti-immunoglobulin-induced caps (61). These
drugs also interfere with Con A-induced mitogenesis (60, 62, 63),
suggesting that interaction of Con A receptors with the microtubular
apparatus of the cell is essential for mitogenesis.

CONCLUSIONS

According to the current model of the cell membrane, proteins,
glycoproteins and glycolipids are embedded in the lipid bilayer
which constitutes the membrane. Some of these components are
protruding out of the plane of the membrane serving as cell surface
receptors. A variety of agents are able to bind to such receptors;
nevertheless, only certain types of ligand—cell interactions result
in lymphocyte stimulation. Such interactions are not dependent on
the specificity of a certain type of receptors, since mitogenic
agents of different specificities are capable of stimulating lymphocytes. In the case of mitogenic interactions, not all binding sites
serve as mitogenic inducers. It seems that the majority of binding
sites or the binding affinity of the ligand to the bulk of the receptors is irrelevant to mitogenic activation. It is rather the

interaction of a mitogenic ligand with only a small fraction of the receptors that is important for stimulation. The current model of lymphocyte activation suggests that multivalent interactions with the appropriate surface receptors result in cross linking of receptors and conformational changes in membrane components. Such perturbation at the outer surface are transmitted through the "carrier" proteins or lipids to cytoskeletal elements inside the membrane, such as microtubules and microfilaments, which then transmit the signal to the cytoplasm and nucleus.

REFERENCES

1. Nowell, P.C. (1960) Cancer Res. 20: 462
2. Lis, H., and Sharon, N. (1977) in "The Antigens," (M. Sela, ed.), Vol. 4, p. 429.
3. Oppenheim, J.J., and Schechter, B. (1976) in "Manual of Clinical Immunology" (N.R. Rose, ed.), Washington: American Soc. for Microbiology, p. 81.
4. Schechter, B., Handzel, Z.T., Altman, Y., Nir, E., and Levin, S. (1976) Clin. Exp. Immunol. 27: 478.
5. Bloom, B.R. (1971) Adv. Immunol. 13: 102.
6. Granger, G.A. (1972) Ser. Haematol. 5(4): 8.
7. Granger, G.A., Daynes, R.A., Runge, P.E., Prieur, A.M., and Jeffes, E.W.B. (1975) in Contemp. Top. Molec. Immunol. 3:205.
8. Andersson, J., and Melchers, F. (1973) Proc. Nat. Acad. Sci. US 70: 416.
9. Melchers, F., and Andersson,. J. (1973) Transplant. Rev. 14: 76.
10. Blease, R.M., Muchmore, A.V., and Nelson, D.L. (1976) in "Mitogens in Immunobiology" (J.J. Oppenheim and D.L. Rosenstreich, eds.), p. 443.
11. Lis, H., Sela, B.A., Sachs, L., and Sharon, N. (1970) Biochim. Biophys. Acta 211: 582.
12. Novogrodsky, A., and Katchalsky, E. (1973) Proc. Nat. Acad. Sci. US 70: 2515.
13. Novogrodsky, A., Lotan, R., Ravid, A., and Sharon, N. (1975) J. Immunol. 115: 1243.
14. Novogrodsky, A. (1976) in "Mitogens in Immunobiology" (J.J. Oppenheim and D.L. Rosentriech, eds.), p. 43.
15. Andersson, J., Sjoberg, O., and Moller, G. (1972) Transpl. Rev. 11: 131.
16. Greaves, M.F., Bauminger, S., and Janossy, G. (1972) Clin. Exp. Immunol. 10: 537.
17. Stobo, J.D., Rosenthal, A.S., and Paul, W.E. (1972) J. Immunol. 108: 1.
18. Boldt, D.H., MacDermott, R.P., and Jordan, E.P. (1975) J. Immunol. 114: 1532.
19. Krug, U., Holtenberg, M.D., and Cuatrecasas, P. (1973) Biochem. Biophys. Res. Comm. 52: 305.

20. Toyoshima, S., and Osawa, T. (1975) J. Biol. Chem. 250: 1657.
21. Presant, C.A., and Kornfeld, S. (1972) J. Biol. Chem. 247: 6937.
22. Weber, T. (1973) Experientia 29: 863.
23. Novogrodsky, A., Biniaminov, M., Ramot, B., and Katchalsky, E. (1972) Blood 40: 311.
24. Kornfeld, S., and Siemers, C. (1974) J. Biol. Chem. 249: 1295.
25. Stein, M.D., Sage, H.J., and Leon, M.A. (1972) Arch. Biophys. Biochem. 150: 412.
26. Betel, I., and Van den Berg, K.J. (1972) Eur. J. Biochem. 30: 571.
27. Inbar, M., Ben-Bassat, H., and Sachs, L. (1973) Int. J. Cancer 12: 93.
28. Allan, D., and Crumpton, M.J. (1973) Exp. Cell Res. 78: 271.
29. Scatchard, G. (1949) Ann. NY Acad. Sci. 51: 660.
30. Prujansky, A., Ravid, A., and Sharon, N. (1978) Biochim. Biophys. Acta 508: 137.
31. Loontiens, F.G., Clegg, R.M., Van Landschoot, A., and Jovin, T.M. (1977) Eur. J. Biochem. 78: 465.
32. Loontiens, F.G., Clegg, R.M., and Jovin, T.M. (1977) Biochemistry 16: 159.
33. Barnett, R.E., Scott, R.E., Furcht, L.T., and Kersey, J.H. (1974) Nature 249: 464.
34. Schechter, B., Lis, H., Lotan, R., Novogrodsky, A., and Sharon, N. (1976) Eur. J. Immunol. 6: 145.
35. Ruddon, R.W., Weisenthal, L.M., Lundeen, D.E., Bessler, W., and Goldstein, I.J. (1974) Proc. Nat. Acad. Sci. US 71: 1848.
36. Prujansky, A., Ravid, A., Lis, H., and Sharon, N. (1978) Biochim. Biophys. Acta, in press.
37. Ravid, A., Novogrodsky, A., and Wilchek, M. (1978) Eur. J. Immunol. 8: 289.
38. Greaves, M.F., and Bauminger, S. (1972) Nature New Biol. 235: 67.
39. Gunther, G.R., Wang, J.L., Yahara, I., Cunningham, B.A., and Edelman, G.M. (1973) Proc. Nat. Acad. Sci US 70: 1012.
40. Sela, B.A., Wang, J.L., and Edelman, G.M. (1975) Proc. Nat. Acad. Sci. US 72: 1127.
41. Beppu, N., Terao, T., and Osawa, T. (1975) J. Biochem. 79: 1113.
42. Fraser, A.R., Heraperly, T.T., Wang, J.L., and Edelman, G.M. (1976) Proc. Nat. Acad. Sci. US 73: 790.
43. Ellner, J.J., Lipski, P.E., and Rosenthal, A.S. (1976) J. Immunol. 116: 876.
44. Lipski, P.E., Ellner, J.J., and Rosenthal, A.S. (1976) J. Immunol. 116: 868.
45. Rosenstreich, D.L., and Wilton, J.M. (1975) in "Immune Recognition" (A.S. Rosenthal, ed.), p. 113.
46. Rosenstreich, D.L., Farrar, J.J., and Dougherty, S. (1976) J. Immunol. 116: 131.
47. Frye, L.D., and Edidin, M. (1970) J. Cell. Sci. 7: 319.
48. Taylor, R.B., Duffus, P.H., Raff, M.C., and de Petris, S. (1971) Nature New Biol. 233: 225.
49. Sharon, N., and Lis, H. (1975) Methods in Membr. Biol. 3: 143.

50. Nicolson, G.L. (1976) Biochim. Biophys. Acta 457: 57; 458: 1.
51. Ahmann, G.B., and Sage, H.J. (1974) Cell. Immunol. 10: 183.
52. Betel, I., and Van den Berg, K.J. (1972) Eur. J. Biochem. 30:571.
53. Ono, M., Maruta, H., and Mizuno, D. (1973) J. Biochem. 73:235.
54. Greaves, M.F., and Janossy, G. (1972) Transpl. Rev. 11: 87.
55. Loor, F. (1973) Eur. J. Immunol. 3: 112.
56. Ahman, G.D., and Sage, H.J. (1974) Cell. Immunol. 13: 407.
57. Inbar, M., Shinitzky, M., and Sachs, L. (1973) J. Mol. Biol. 81: 245.
58. Karnovsky, M.J., and Unanue, E.R. (1975) Fed. Proc. 32: 57.
59. Cunningham, B.A., Wang, J.L., Gunther, G.R., Reeke, G.N., Jr. and Becker, J.W. (1974) in "Cellular Selection and Regulation in the Immune Response" (G.M. Edelman, ed.), pp. 177.
60. Edelman, G.M., Yahara, I., and Wang, J.L. (1975) Proc. Nat. Acad. Sci. US 70: 1442.
61. Yahara, I. and Edelman, G.M. (1973) Nature (Lond.) 246: 152.
62. Medrano, E., Piras, R., and Mordoh, J. (1974) Exp. Cell Res. 86: 295.
63. Wang, J.L., Gunther, G.R., and Edelman, G.M. (1975) J. Cell Biol. 66: 128.

DNA REPLICATION UNITS IN EUKARYOTES

Francesco Amaldi

Centro di Studio per gli Acidi Nucleici, C.N.R.
Istituto di Fisiologia Generale
Università di Roma

The eukaryotic chromosome contains a single DNA molecule whose length ranges, in different organisms and
different individual chromosomes, from few mm up to the
order of 1 m. Very simple calculations show that a DNA
molecule of such astonishing size cannot replicate starting from a unique origin and proceeding at one or two
(bidirectional) replication forks as it occurs in prokaryotic chromosomes and in the small DNA molecules of
plasmids, viruses etc.

The presence of many independent DNA replicating
units (or "replicons") within a single eukaryotic chromosome has been experimentally demonstrated in 1959 by
Taylor (1). After a 10 min pulse with ^{3}H-thymidine of
Chinese hamster cells he reports that in cells, which
had then reached the metaphase, some of the larger chromosomes showed, after autoradiography, several labelled
regions in a single arm. In the following years similar
results have been obtained by other authors in different
systems, animals and plants, supporting the view of multiple simultaneous DNA synthesis sites.

The magnitude of the number of sites of DNA replication per chromosome had not become apparent until the
DNA fiber autoradiography technique, first developed by

Cairns (2), had been applied to mammalian DNA replication by Cairns himself (3) and by Huberman and Riggs (4). This technique provides a much greater resolution than is attainable when metaphase chromosomes in whole cells are used.

Following this technique cells are labelled with ^{3}H-thymidine at the highest possible specific radioactivity, then gently lysed and, in one of a number of ways, the cellular DNA is stretched out on a flat surface and autoradiographed. This method enables single extended DNA molecules to be visualized under light microscope.

By this method Cairns (3) has observed autoradiograms of about 10-30 μm and of 50-100 μm after pulse experiments of 45 and 180 min, respectively (human lymphocyte cells); thus the **average growth rate of** the newly synthesized **sections of DNA was** 0.5 μm/min or less. At this growth rate, at least 100 (probably many more) replication units may be required to replicate all the DNA on a chromatid within the six hours needed by human cells for DNA synthesis (S-phase). Moreover, Cairns suggested on the basis of the autoradiograms observed, that replication units were tandemly joined to form long DNA fibers.

More information is provided by Huberman and Riggs in the paper published in 1968 (5). Their autoradiographic studies of DNA from Chinese hamster cells and HeLa cells pulse labelled with ^{3}H-thymidine show very clearly that many replicons are joined together in series as proposed by Cairns (Fig. 1a). They show, moreover, that each replicon involves two growing points that proceed in opposite directions from a common origin. Evidence for this bi-directional synthesis of DNA (which has been demonstrated in this system earlier than in prokaryotic systems) comes from pulse-chase experiments which produced autoradiographs showing heavily labelled segments flanked on both sides by diminishing grain-density gradients (Fig. 1b). The distances between neighbouring replication origins vary and range mostly between 15 and 60 μm. Neighbouring origins may start replication at different times. (This fact is connected with the al-

Fig. 1. Bidirectional DNA replication in a mammalian chromosome. (a) Autoradiographs of Chinese hamster DNA molecules from cells which have briefly been labeled with tritiated thymidine. The tandem arrays of exposed grains indicate the existence of several replication points in the portion of the DNA fibers under view. (b) Tandem arrays seen after a pulse exposure to label followed by a chase period in nonradioactive medium. Here the grain declines from the center to the ends, suggesting that the growing points move in opposite directions. Reproduced from Huberman and Riggs (5).

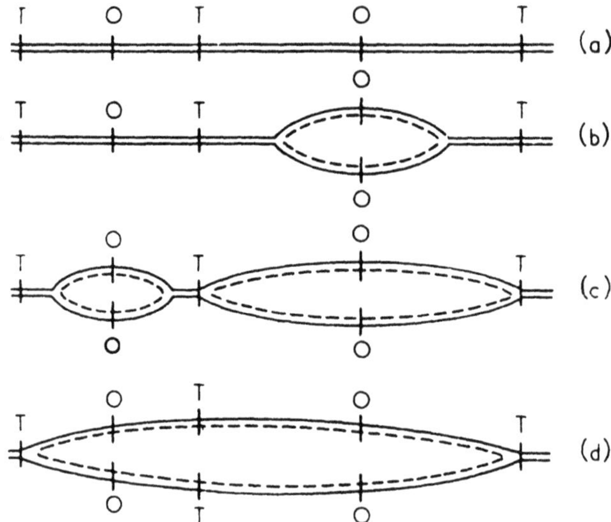

Fig. 2. The bi-directional model for DNA replication of Huberman and Riggs (5). Each pair of horizontal lines represents a section of a double helical DNA molecule containing two polynucleotide chains (———, parental chain; ————, newly synthesized chain). The short vertical lines represents positions of origins (O) and termini (T). The diagrams represent different stages in the replication of two adjacent replication units.
(a) Prior to replication
(b) Replication started in right-hand replication unit
(c) Replication started in left-hand unit and completed at termini of right-hand unit
(d) Replication completed in both units; sister double helices separated at common terminus.

ready known asynchrony of DNA replication for which some
chromosome regions replicate at different times from
others. The pattern of asynchrony is reproducible among
different cells of the same type.)

From the origin, DNA replication proceeds, at 37°C,
at about 2 μm per min at each growing point. A simple
calculation will show that a mammalian cell may have
about 10,000 such replication units. The model proposed
by Huberman and Riggs is shown in Fig. 2.

After the mentioned fundamental work of Huberman and
Riggs on Chinese hamster and HeLa cells, several other
authors have utilized this technique to extend the observations to other systems and to study the dependence
of the two parameters, rate of fork movement (R = rate)
and replicon size (ID = initiation distance), in various conditions. Thus the general validity of the Huberman and Riggs model has been demonstrated on vertebrates
other than mammals (chicken (6) and amphibian (7)) in
Drosophila (8) and even in plants, Pisum sativum (9).
These and several other studies have shown that replication units are quite variable in size according to the
species and to the cells or tissues considered.

The rate of fork movement (R) and replicon size (ID)
have been studied by Callan (7) in relation to genome
size. This author has studied two amphibians, Xenopus
and Triturus, the latter having about 10 times more DNA
per cell than the former, with however a comparable number of chromosomes. Both ID and R appear to be larger
in Triturus than in Xenopus; the results have been obtained, though, comparing two different cell types (cultures of an established cell line of Xenopus, primary
cell cultures from Triturus liver). The author also suggests that the very long S-phase of Triturus spermatocytes
is due to a gross reduction in the number of initiation
points (larger ID) rather than a change of R.

An increase of R has been observed during the S phase
of Chinese hamster cultured cells, where R increases three
times from the early to late S-phase while ID remains about

the same (10). Similar results have been obtained in the
plant <u>Pisum sativum</u> (9).

It has also been shown that DNA synthesis inhibition
by FUdR reduces both R and ID but not to the same extent
in different mammalian cell lines (11). As for the effect of protein synthesis inhibition it has been observed
that a puromycin treatment reduces ID but not R in Chinese hamster cells and in mouse cells (12, 13), while
studying several mammalian cell lines, Stimac <u>et al</u>. (14)
observed a reduction of both R (immediately after addition of the inhibition) and ID (somewhat later). These
authors have carefully discussed a relationship of protein synthesis, the overall DNA synthesis, ID and R in
relation to the mechanism of coupling between DNA synthesis and protein synthesis.

Hand and Tamm (15) have studied the initiation of
DNA replication after infection of mammalian cells by
Reoviruses and have concluded that ID is reduced by viral
infection. A similar reduction of ID has been demon-
strated in Chinese hamster cells after transformation by
SV40 (16).

Another problem which has seemed possible to be approached by the DNA fiber autoradiography technique is
the reproducibility of the replicon origins. The question was "Are the origins specific sites along the DNA
molecules, thus always the same in a given cell type?"
An attempt to answer this question has been done (17, 18)
by pulse labelling at two subsequent cell cycles Chinese
hamster cells growing in synchrony. Ideally, two different isotopes should have been used in order to distinguish between the autoradiograms corresponding to the
first and to the second pulse. This being impossible
for technical reasons, both pulses were done with ^{3}H-
thymidine but for different times (10 min and 25 min for
the first and the second pulse, respectively). By doing
so, we had expected to be able to distinguish the autoradiograms on the basis of their length and to see if
they are centered with each other, indicating reproducible origins or not. Unexpectedly, we had found that
the two labellings rarely overlap but when they do they

appear to be centered with each other.

These results had suggested that the origins are reproducible but the time order of their activation is not strictly fixed (of course this has nothing to do with the gross reproducibility of the pattern of asynchrony of DNA replication which is clearly demonstrated).

Although DNA autoradiography is still used and can in fact be useful in the study of certain problems, it cannot tell us much more on the structure and organization of chromosomal DNA replicons.

Electron microscopy, which could have provided the higher resolution required, has been on the contrary exploited only lately to study the organization of eukaryotic replicons mainly because of two disadvantages it has with respect to DNA fiber autoradiography: 1) the size of replicons is large, about 30 μm in mammals for instance, so that it is difficult to observe at the electron microscope DNA fragments with length corresponding to more than one replicon; even when such long molecules are found they are often untangled and difficult to be interpreted; 2) in a DNA preparation, the frequency of replicating molecules is usually extremely low, and long and tedious scanning of the grids is required to find a replication form among innumerable simple linear molecules; by DNA fiber autoradiography, on the contrary, we observe in the preparation only the replicating molecules while the non-replicating ones, by definition, do not appear at all.

To bypass these two drawbacks of the electron microscopy technique, Blumenthal *et al*. (8) and Kriegstein & Hogness (19) have studied by this method the cleavage nuclei of *Drosophila melanogaster*. At 24°C the cleavage nuclei devide every 9.6 min in the syncytium of the egg for a period of about 2 h after fertilization. Since interphase occupies 3.4 min of this doubling time, chromosomal DNA molecules are presumably replicated within this short period. It is obvious that in such material, the frequency of replicating DNA molecules expected to be ob-

served at the electron microscope should be high enough.
The other advantage of this system, with respect to the
more classical culture of mammalian cells, is that the
size of replicons (ID) is in _Drosophila_ much shorter, as
the authors themselves have shown (5-10 µm in comparison
with 30-60 µm in mammalian cells); thus multireplicon molecules can easily be observed and followed at the electron microscope.

When these authors had examined DNA isolated from
cleavage nuclei they had observed molecules containing
multiple eye forms (fig. 3). That these eye forms result from DNA replication is indicated by the facts
that the branches of an eye have the same length and are
mainly double-stranded; moreover, partial denaturation
of the two segments of an eye exhibits the same denaturation map. The authors have presented quantitative analysis of the material; eye lengths and eye-to-eye distance. They have also observed that most forks have
single-stranded regions associated with them. These single-stranded containing forms have a configuration and
an orientation predicted from known features of DNA replication forks.

All these observations are in perfect agreement with
the Huberman and Riggs model for DNA replication, model
which had been derived from DNA-fiber autoradiography.
Thus it had seemed that the higher resolution of electron
microscopy had not contributed a new insight in the process of DNA replication but simply confirmed our previous
concepts. It is probably for these reasons that for a
few years this approach had almost been abandoned. Few
papers had appeared extending this observation to other
systems; for instance, long eyes had been described in
replicating DNA from yeast (20) and from _Cochliomyia_
homnivorax (21).

Something new is described in 1976 by Virginia Zakian
(22). She had studied at the electron microscope, replicating main band DNA and satellite DNA from _Drosophila_
virilis cleavage nuclei. She reports two classes of
presumptive replicating molecules with more than one site

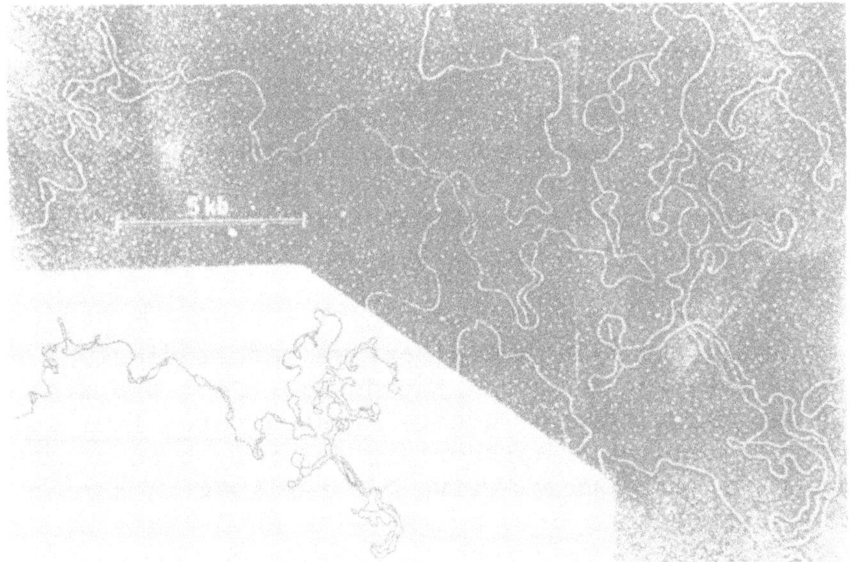

Fig. 3. Electron micrograph of replicating DNA from
<u>Drosophila melanogaster</u>. The portion of chromosomal
molecule shown here is 119 Kb in length and contains
23 eye forms. This electron micrograph is reproduced
from Kriegstein and Hogness (19), which should be con-
sulted for the method and conditions.

of replication in main band DNA. The first class containing one or more clusters of small eyes (all eyes less than or equal to 900 bases in size), the second class comprising molecules with eyes of larger sizes as those described by Blumenthal et al. (8). The author has interpreted these results as follows: replication in main band DNA is initiated by activation of 2 to 12 closely-spaced origins; these clusters of origins are spaced out at regular intervals along the chromosomes and eventually fuse to produce larger eyes. The presence of a tandem array of small "bubbles" had been previously reported in replicating mammalian DNA in 1973 by Bick (23).

The model of Huberman & Riggs should thus be modified only since multiple closely-spaced origins are present in each replicon; these multiple origins, of course, cannot have been resolved by the DNA-fiber autoradiographic technique.

More recently, in our laboratory, we had studied DNA replication with the electron microscope (24); the system we had chosen is the sea urchin Paracentrotus lividus. This material is most appropriate for this type of study, since it has a very active DNA replication during cleavage (similar to Drosophila) with an S-phase of few minutes; moreover, synchronous cleavage can be easily obtained and the amount of material is no problem (as is the case with Drosophila).

We had studied mainly DNA from embryos at the third S-phase, grown generally at 15°C and occasionally at 10°C, gastrulae and adult tissues. The electron microscopic analysis of DNA purified from third S-phase embryos had revealed a number of typical structures. The most characteristic one is represented by very small eyes (microbubbles) mostly tandemly grouped in clusters consisting of an average of 7 microbubbles up to a maximum of 20. We had often found more than one cluster on the same DNA molecule. Average length of microbubbles is 0.3 Kb and their center-to-center distance within each cluster is variable with an average of 0.7 Kb. The distance between clusters is of the order of few μm (Fig. 4).

The microbubbles are unstable structures as they disappear with time during storage at 4°C or in destabilizing conditions (high temperature or in presence of formamide). These observations, together with S1 nuclease sensitivity and RNase resistance of microbubbles, suggest that they are partially single stranded structures; snap back of parental strands and extrusion of newly synthesized DNA segments (branch migration) might account for the instability of microbubbles. A second typical structure of DNA from early cleavage embryos is the abundance of single strands (S1 nuclease sensitive and RNase resistant) both free and continuous with double-stranded DNA (Fig. 5). Single strands and clusters of microbubbles are significantly associated on the same molecules.

Similar results are also obtained when DNA is purified from a later developmental stage and from adult tissues. The frequency of the described structures is, however, lower, and the relative amount of single strands with respect to microbubbles is increased.

The most intriguing aspect of our observations, which had led us to carry out a number of control experiments, is the complete absence of long eyes expected to result from the confluence of growing microbubbles.

Let us summarize the results obtained in the various systems by different authors. Two of the few eukaryotic species studied up to now have revealed the presence of growing long eyes, namely yeast (20) and _Cochliomyia homnivorax_ (21). On the other hand, we have observed in _Paracentrotus lividus_ the presence of only clusters of microbubbles and single-stranded containing structures (24). We have also preliminary data showing a qualitatively similar pattern in a variety of animals and plant systems: chick embryos; _Xenopus laevis_ embryos, larvae and cultured cells; goldfish adult tissues; Chinese hamster cultured cells, Vicia faba root tips (Buongiorno Nardelli, Carnevali, Baldari & Micheli, unpublished

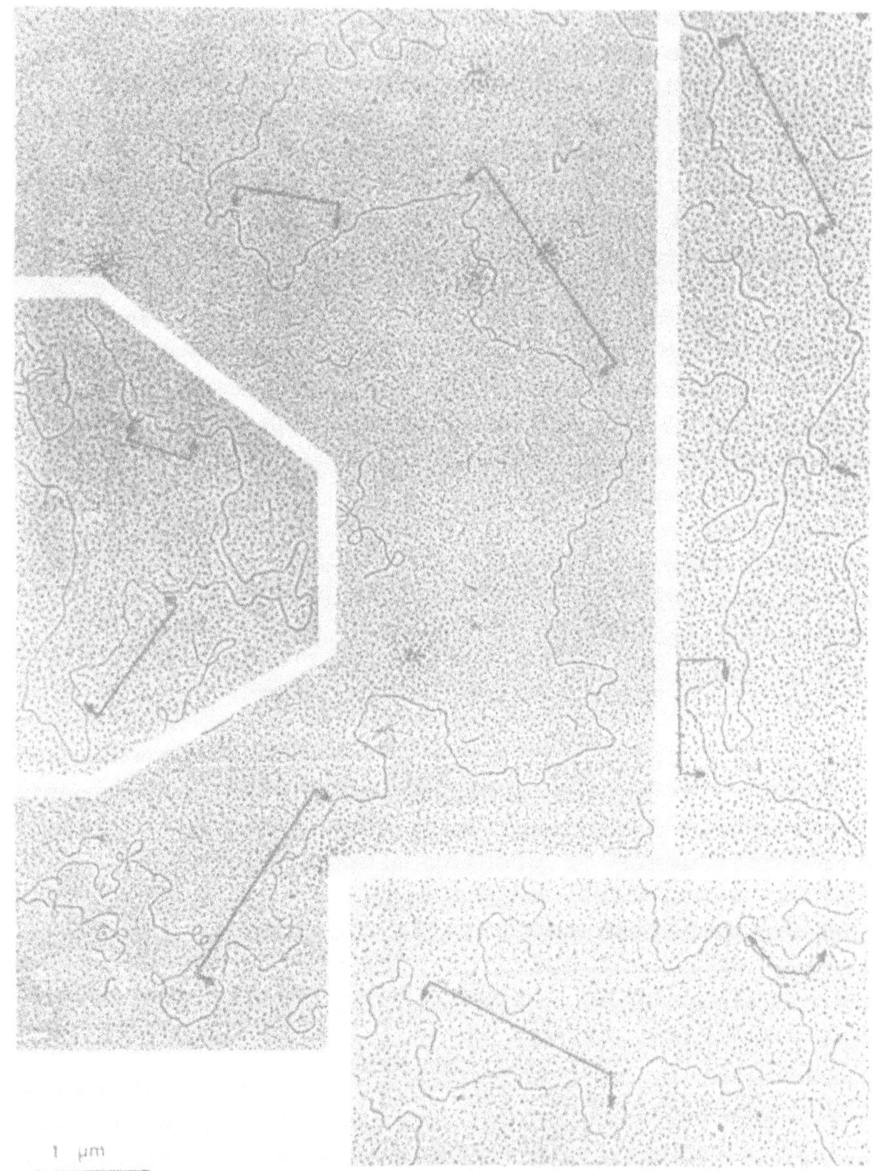

Fig. 4. Clusters of microbubbles observed at the electron microscope in replicating DNA from *Paracentrotus lividus* early cleavage embryos. Reproduced from Baldari, Amaldi and Buongiorno-Nardelli 24).

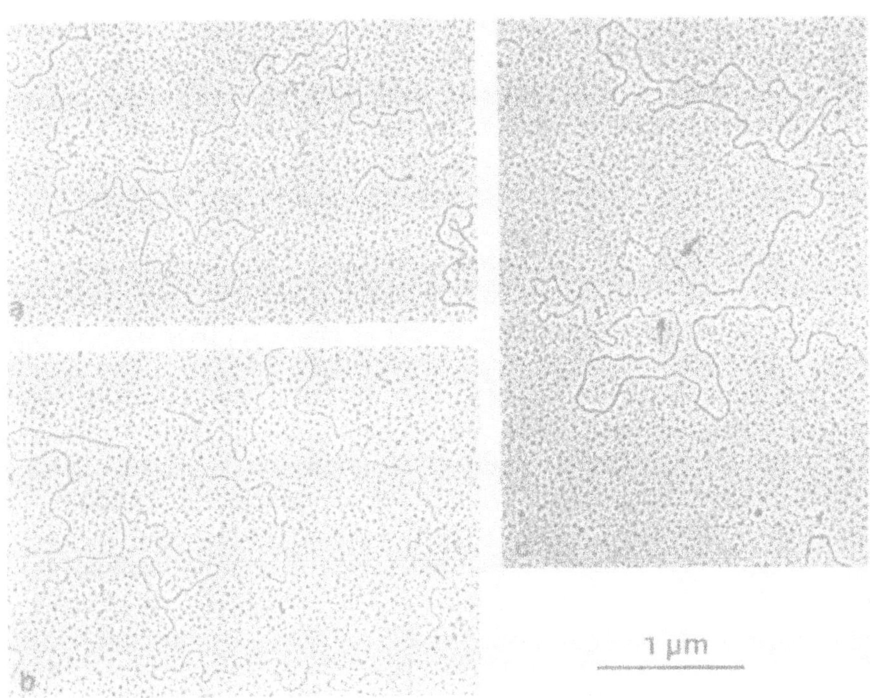

Fig. 5. Electron micrographs of replicating DNA from
Paracentrotus lividus early cleavage embryos. In a)
and b) free single-stranded molecules are shown. In
c) a double-stranded molecule with a single-stranded
gap can be observed. Due to different PT shadowings,
the absolute thickness of double- and single-stranded
DNA can vary in different micrographs. Reproduced
from Baldari, Amaldi and Buongiorno Nardelli (24)

data). As _for Drosophila_, Blumenthal _et al_. (8) have
observed in _D. melanogaster_ only long eyes, while Zackian
(22) in _D. virilis_ has found both long eyes and clusters
of microbubbles. In a detailed study not yet published,
we have observed in _D. melanogaster_ both long eyes and
clusters of microbubbles plus partially single-stranded
molecules; but, what is more important, we could show
that the pattern observed is stage-dependent, long eyes
being found only in DNA prepared from very early cleavage
nuclei, whereas microbubbles and single-stranded containing structures are present in DNA prepared from later
stages. A similar situation seems to occur in the sea-
urchin _Arbacia_ (Stambrook, personal communication).

The presence of microbubbles and single-stranded
structures and the absence of long eyes in DNA from actively dividing cells may be explained as follows. Chromosome duplication might occur in two steps. The first
would consist in the initiation of DNA replication at
multiple origins per each replicon (22). This synthesis
would stop soon, for instance because of topical constrains, resulting thus in clusters of microbubbles.
These would then be structures analogous to the D-loops
described in mitochondrial DNA (25). In the subsequent
step the removal of constrains, for example by single-
strand nicks, would allow DNA synthesis to resume. This
synthesis would proceed asymmetrically, that is on one
strand only, the other being replicated later, accounting
thus for the abundance of both single-strand structures
and the occasional presence of asymmetrical linear forks.
The absence of long eyes would be the result of the removal of proteins involved in maintaining integrity of parental strands where nicks had been introduced. It is,
however, difficult to concile this view with the bi-direc-
tionality of replication, as seen in several eukaryotic
systems by DNA fiber autoradiography (which involves an
SDS treatment), and does not account for the different
DNA structures (microbubbles and long eyes) observed
either within the same species at different stages of
development (see for instance _Drosophila melanogaster_)
or between closely related species (such as _Paracentrotus_
and _Arbacia_).

We hence propose an interpretation of the results
obtained both by DNA fiber autoradiography and electron
microscopy in different systems, based on the assumption
that the process of DNA replication occurs in three se-
quential and uncoupled steps: a) separation of parental
strands by "active unwinding proteins", starting at replicon origins and proceeding bi-directionally; b) discontinuous DNA synthesis (Okazaki fragments) on both the
previously unwound strands; c) completion and ligation
of fragments. The presence of either microbubbles or
long eyes in the different systems analyzed can be interpreted if the rate of the process of DNA replication
is limited by any one of the three steps. For example,
if parental strand separation is rate-limiting, long
eyes are observed at the electron microscope as, for instance, in Drosophila early cleavage nuclei (8, 19). If,
on the other hand, discontinuous synthesis is rate limiting, leaving parental strand separation to proceed ahead,
single-stranded DNA will appear where reassociation of
parental strands following DNA extraction is not possible
because of breakages. Finally, if completion and ligation of newly synthesized fragments lag behind, snap-back
of parental strands during DNA extraction will result in
clusters of microbubbles where Okazaki fragments had been
synthesized. It is noteworthy that the size of microbubbles we have described in sea urchin DNA compares well
with the average length of Okazaki fragments in eukaryotes
(26). In conclusion, microbubbles would be artifacts due
to removal of unwinding proteins. This tentative explanation does not contrast with the model of bi-directional
chromosomal replication based on DNA fiber autoradiography
experiments (5). It would imply though a discontinuous
DNA synthesis on both strands; moreover, Okazaki fragments
would be synthesized not sequentially, leaving gaps to be
filled later, as proposed by Hand on the basis of incorporation experiments (personal communication).

The interpretation proposed here can now be tested by
analyzing, by means of electron microscopy, chromosome
duplication at a higher organization level, namely the
DNA associated with proteins in the chromatin structure,
as has recently been performed by McKnight and Miller (27)

on <u>Drosophila</u> <u>melanogaster</u>. It should also be possible to affect the three steps of chromosome replication with appropriate drugs.

The limitations of all these studies rest on the fact that observations have involved the whole genome refer- ring to an "average" replicon and an "average" origin. No one can be satisfied with this at present days; pre- cise information is now necessary on the structure of a given replicon and its origin including their nucleotide sequences. In fact, recently a number of DNA replication origins have been identified and in several cases their nucleotide sequences determined. Up to now, however, they refer to prokaryotic replicons (<u>Escherichia</u> <u>coli</u> chromosomes, plasmids and phages) and few eukaryotic non- chromosomal replicons (mitochondrial DNA, viruses). Brief- ly, it has been shown that the sequence involved in the origin of DNA replication is quite long, more than a pro- motor for instance, and that it is a specific sequence that can vary in different cases, but seems to share a similar possibility for secondary structures.

It is probable that, in the next future, efforts will be made to carry on similar detailed studied on eukaryotic chromosomal replicons. In our laboratory, for instance, work is in progress to identify the origin of replication of the multiple copies of the rRNA gene in <u>Xenopus</u> <u>laevis</u>. This gene is easily isolated by CsCl density gradients and has a fairly well known structure. The study is car- ried out by electron microscpy of replicating forms (clus- ters of microbubbles) in the rDNA purified from <u>X. laevis</u> larval cells and by radioactive pulse labelling of the origins in synchronized <u>X. laevis</u> cultured cells. In both cases digestions with restriction enzymes allow the localization of the origin. The results obtained up to now indicate that each rDNA repeating unit has its own origin of replication and that this is localized within the non-transcribed spacer (Bozzoni, Baldari, Amaldi & Buongiorno Nardelli, manuscript in preparation).

Another direction for present and future research is the study of the relevance of chromatin structure for euka- ryotic DNA replication.

1) Taylor, J.H. (1959). In "Proc. 10th Intern. Congr. Genet., Toronto", Toronto University Press, vol. 4, pp. 63-78.
2) Cairns, J. (1962). J. Mol. Biol. 4, 407-409.
3) Cairns, J. (1966). J. Mol. Biol. 15, 372-373.
4) Huberman, J.A. & Riggs, A.D. (1966). Proc. Nat. Acad. Sci. U.S. 55, 599-606.
5) Huberman, J.A. & Riggs, A.D. (1968). J. Mol. Biol. 32, 327-341.
6) McFarlane, P.W. & Callan, H.G. (1973). J. Cell Sci. 13, 821-839.
7) Callan, H.G. (1972). Proc. Royal Soc. London B. 181, 19-41.
8) Blumenthal, A.B., Kriegstein, H.J. & Hogness, D.S. (1973). Cold Spring Harbor Symp. Quant. Biol. 38, 205-223.
9) Van't Hof, J. (1976). Exp. Cell Res. 103, 395-403.
10) Housman, D. & Huberman, J.A. (1975) J. Mol. Biol. 94, 173-181.
11) Ockey, C.H. & Saffhill, R. (1976). Exp. Cell Res. 103, 361-373.
12) Hori, T. & Lark, K.G. (1973). J. Mol. Biol. 77, 391-404.
13) Hand, R. & Tamm, I. (1972). Virology 47, 331-337.
14) Stimac, E., Housman, D. & Huberman, J.A. (1977). J. Mol. Biol. 115, 485-511.
15) Hand, R. & Tamm, I. (1974). J. Mol. Biol. 82, 175-183.
16) Martin, R.G. & Oppenheim, A. (1977). Cell 11, 859-869.
17) Amaldi, F., Carnevali, F., Leoni, L. & Mariotti, D. (1972). Exp. Cell Res. 74, 367-374.
18) Amaldi, F., Buongiorno Nardelli, M., Carnevali, F., Leoni, L., Mariotti, D. & Pomponi, M. (1973). Exp. Cell Res. 80, 79-87.
19) Kriegstein, H.J. & Hogness, D.S. (1974). Proc. Nat. Acad. Sci. U.S. 71, 135-139.
20) Newlon, C.S. Petes, T.D., Hereford, L.M. & Fangman, W.L. (1974). Nature 247, 32-35.
21) Lee, C.S. & Pavan, C. (1974). Chromosoma 47, 429-437.
22) Zakian, V.A. (1976). J. Mol. Biol. 108, 305-331.
23) Bick, M.D. (1973). Ninth Intern. Congress of Bioche-

mistry, Stockholm, p. 135.
24) Baldari, C.T., Amaldi, F. & Buongiorno Nardelli, M. (1978). Cell 15, 1095-1107.
25) Kasamatsu, H., Robberson, D.L. & Vinograd, J. (1971). Proc. Nat. Acad. Sci. U.S. 68, 2252-2257.
26) Nuzzo, F., Brega, A. & Falaschi, A. (1970). Proc. Nat. Acad. Sci. U.S. 65, 1017-1024.
27) McKnight, S.L. & Miller, O.L. (1977). Cell 12, 795-804.

FUNCTIONS OF DNA POLYMERASES α, β AND γ IN DNA
REPLICATION AND REPAIR

Miria Stefanini, Anna I. Scovassi, Umberto Bertazzoni[*]

Laboratorio di Genetica Biochimica ed Evoluzionistica
del Consiglio Nazionale delle Ricerche
Via S. Epifanio 14
27100 Pavia, Italy

The cells of vertebrate organisms contain three distinct
DNA polymerases which have been designed α, β and γ -polymer-
ases (1). The information concerning the physical, chemical and
catalytical properties of these enzymes has progressed considerably during the past few years so that the distinction of their
activities is easily obtained by utilizing the differences in molecular weight, sensitivity to inhibitors, chromatographic elution,
ability to copy various templates (reviews: 2-8). The general
properties of the three enzymes are summarized in Table I.

The DNA polymerase α represents the major activity in dividing cells; it is a high molecular weight enzyme made by a
single polypeptide chain (2). It belongs to the nucleus but is not
usually found within the nucleus following aqueous extraction.
Bollum has recently observed that, using antibodies against
highly purified α-polymerase, the fluorescence is found mainly
in the perinuclear region (9). The enzyme has not been obtained
yet in the homogenous form. It is characterized by its sensitivity to high salt and to the sulphydryl blocking agent NEM. According to recent reports, it is also inhibited by the antibiotic
aphidicolin (10) whereas it is not affected by the nucleotide
analog dideoxy-TTP (11).

The β-polymerase is a low molecular weight enzyme which

* EMBO invited speaker.

Table I

Properties of vertebrate DNA polymerases

Properties	α	β	γ
Molecular weight	130000	45000	130000
Cellular localization	perinuclear	nuclear	mitoch.; nuclear
% of total activity (growing cells)	70 - 80	10 - 15	5 - 10
Polypeptide chain	single	single	
Specific activity (U/mg)	100000	200000	60000
Optimal pH	7.2	8.6	8.0
Isoelectric point	5.5	9 - 9.4	5.4 - 6.1
High ionic strenght effect	inhibition	stimulation	stimulation
Effect of N.E.M. (N-Ethyl Maleimide)	inhibition	no effect	inhibition
Effect of dideoxy TTP (ddTTP)	no effect	inhibition	inhibition
Effect of aphidicolin	inhibition	no effect	no effect
Associated DNases	no	no	no
K_m for dNTP's (μM)	2 - 12	8 - 12	0.2 - 0.6

Table I / contd.

Properties	α	β	γ
Effect of Helix Destabilizing Proteins	stimulation	no effect	no effect
Misincorporation frequency	10^{-4}	10^{-4}	10^{-4}
Ability to bypass pyrimidine dimers	yes	yes	yes
Activity with template-primer complexes:			
activated DNA	yes	yes	poor
$(rA)_n \cdot (dT)\,\overline{15}$	no	yes	yes
$(rA)_n \cdot (dT)\,\overline{15}$ + K phosphate	no	no	yes
RNA-primed DNA	yes	no	no
oligo dT primed mRNA	no	no	no
Number of molecules/cell (estimated)	25000	5000	

is found in the nucleus by using the standard fractionation techniques. It is the only DNA polymerase which has been purified
to homogeneity (12, 13). It is characterized by having an alkaline
isoelectric point and by being completely insensitive to NEM (2).
The response of β-polymerase to NEM, aphidicolin (10) and
ddTTP (11) is just the opposite of what is found for the α-polymerase, permitting a clear distinction of the two enzymatic activities. The β-enzyme is found in all multicellular animals but
it is absent in bacteria, plants and protozoa (14).

DNA polymerase γ is a high molecular weight enzyme and
represents a minor constituent of total cell polymerases. It is
an acidic protein and is characterized by being sensitive to NEM
and responding preferentially to the homopolymer system composed of a ribotemplate and a deoxyprimer. For this reason the
γ-polymerase has been mistaken in the past for a reverse transcriptase. The distinction from this enzyme is best made on
copying mRNA and from β-polymerase, which uses also poly(A)-
oligo(dT), by inhibiting it with K phosphate buffer which stimulates γ-polymerase activity (15). Concerning the in vivo localization, it has been recently found that the mitochondrial DNA
polymerase, considered before as a distinct enzyme, is undis-
stinguishable from the γ-polymerase (16-19). At present the
existence and the possible function of a nuclear γ-polymerase
remains to be defined.

The understanding of the respective function of the three
DNA polymerases has been made difficult by the lack of pertinent
mutants in eukaryotic cells. However the availability of specific
assays for each polymerase, which allow their quantitative measurement in crude cell extracts, has prompted a number of physiological experiments aimed at the identification of the respective roles of the three DNA polymerases.

We shall try to consider in detail the evidence accumulated
up to now about this problem and to draw tentative conclusions.

DNA polymerase function in DNA replication

The first physiological experiments demonstrating variations of DNA polymerase activities in mammalian systems were
performed about twenty years ago, by showing a striking increase of overall DNA polymerase activity in regenerated rat

liver (20). Similar evidence was obtained using other systems, as for instance stimulated human lymphocytes, where the rise in DNA polymerase activity is parallel to the DNA synthesis rate (21-24).

The first indication that the high molecular weight enzyme (named later α-polymerase) was indeed correlated with DNA replication came from the work of Chang and Bollum. By using different systems (25-27) they invariably demonstrated that the level of α-polymerase responds to the variations of the rate of DNA synthesis. Conversely, the level of the low molecular weight enzyme (β-polymerase) did not change significantly in these different physiological conditions. Similar results were obtained in stimulated spleen cells (2, 28, 29).

Yet in synchronized cultured cells the rise in α-polymerase during S-phase is not striking: in fact the level of the three DNA polymerases is already high before DNA synthesis starts and the cell is committed to double this amount before division (30, 31, V. Hitchins person. comm.). Hübsher et al. (32) have followed the levels of the three DNA polymerases during perinatal development of rat neurons and have shown that α-activity drops sharply from a high level in foetal life to an undetectable value after two-three weeks of postnatal age; this loss in α-polymerase correlates also very well with the decline in mitotic activity. In contrast, the activities of β- and γ-polymerases remain constant during the whole period of tissue differentiation.

Recently different laboratories have reported that α-polymerase is the major DNA polymerase associated with replicating SV40 chromosomes(11, 33,34). In addition Edenberg et al. (11) have shown that SV40 DNA synthesis is resistant to ddTTP and that α-polymerase is the only one of the three mammalian DNA polymerases which is resistant to this nucleotide analog, thus implying that this enzyme is responsible for all phases of the SV40 DNA replication. In order to understand if this conclusion could be extended to cellular DNA synthesis in uninfected cells, Waqar et al. (35) tested the effect of ddTTP on an _in vitro_ synthesizing system from HeLa cells: elongation of continuous strands and initiation, elongation and joining of Okazaki pieces, (i.e. the main aspects of the polymerizing reaction in semiconservative DNA replication) appeared to proceed in normal fashion even at high concentration of the drug. Similar conclusions are reached when dividing cells are treated with

aphidicolin, an antibiotic which specifically inhibits α-polymerase (10). In this case the mitotic division of sea urchin eggs is
blocked, whereas no effect is found on meiotic maturational division in starfish oocytes, a process which is not dependent on DNA
replication (36).

All these observations, when added to the bulk of data gar-
nered from the studies of the properties of the α-polymerase
molecule [polymerization rate which is close to the known chain
growth in vivo (6); unique ability to use RNA-primed-natural DNA
as template (37), as required for DNA replication (38-40); stimulation by helix destabilizing proteins (41-43)] strongly indicate
that DNA polymerase α is mainly responsible for the replication of nuclear DNA.

On the contrary the β-polymerase, as mentioned before,
does not show any direct correlation with the increase in DNA
synthesis, is not associated with replicating SV40 chromosomes
and is inhibited by ddTTP which has no effect on DNA replication.

As far as DNA polymerase γ is concerned, the recent
finding that this enzyme is the only polymerase present in the
mitochondria (16-19) is a convincing argument for its role in
mitochondrial DNA synthesis. A more direct demonstration has
been obtained by Hübscher et al. by showing that in rat synapto-
somes, permeabilized to incorporate radioactive nucleotide
precursors, the newly synthesized DNA arises only by replication (44).

DNA polymerase function in DNA repair

The assignment of functional roles to the three DNA polymerases in the process of DNA repair has been largely specula-
tive for a long time. The fact that the level of β-polymerase
does not correlate with DNA replication rate and remains constant in various physiological conditions, has been taken as an
indication that this enzyme could be related to repair type synthesis. Stronger evidence for its function comes from our work
with PHA stimulated human lymphocytes (45) where variations of
β-polymerase, as opposed to the α-polymerase, do not
parallel the increase in DNA synthesis rate. In fact the β-polymerase reaches its maximum at late times of stimulation in

correspondence to a second increase of DNA polymerase activity (46) when DNA replication rate and α-polymerase activity are levelling off and in coincidence with a peak in the capacity to perform repair synthesis following UV-irradiation (47).

This correlation suggests that indeed the β-enzyme is involved in DNA repair; it is possible however that its increase at late stimulation times could be related to recombination processes which may conceivably be required for the production of immunoglobulins. Furthermore, the β-enzyme is the only polymerase found in mature spermatocytes, where chromosome replication is absent but recombinational events are actively accompanying the meiotic process (48). Indirect evidence of β-polymerase implication in DNA repair-type synthesis is found also in blood granulocytes where both UV-induced DNA synthesis (49) and β-polymerase levels are very low (50).

Similarly, addition of ddTTP, an inhibitor of β-polymerase, to cultures of primary hepatocytes treated with carcinogens, induces a decrease in the resynthesis step of excision repair (51).

Korn and associates have recently reported that β-polymerase is capable to repair with the same efficiency both small and long gaps in DNA whereas α-polymerase binds only to long gaps and in a position far from the 3'OH primer (52). This observation could also explain the recent finding by Coetzee et al. (53) that treatment of DNA with bleomycin in vitro induces a selective increase of β-polymerase activity at the sites of the breaks.

In an attempt to understand further the role of the DNA polymerases in DNA repair we have tried a different approach; we have determined the levels and the sedimentation properties of the three DNA polymerases in fibroblasts from patients with inherited diseases affecting DNA repair processes (reviews: 54, 55) such as Xeroderma Pigmentosum, Fanconi's anemia, Ataxia telangiectasia, Bloom's syndrome, Progeria and Werner's syndrome. No quantitative nor qualitative significant changes were found with respect to control fibroblasts (56, 57). Similar results were obtained in Xeroderma Pigmentosum cells by Parker and Lieberman (58). This could mean that the deficient molecule is not a DNA polymerase, but it cannot be excluded that other factors influencing DNA polymerizing activity are affected.

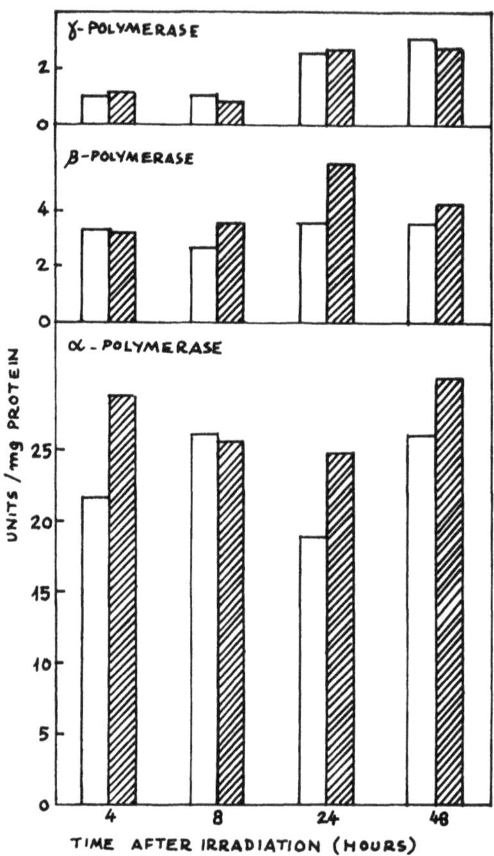

<u>Figure 1.</u>   Levels of DNA polymerases α , β   and γ in EUE
cells UV-irradiated with 6 J/m^2.
The cells were sampled out 4, 8, 24, 48 hours after irradiation and specific activities of the three DNA polymerases
determined on crude extracts as previously described (17).
Non irradiated samples have been pooled at the different times
and used as controls. One unit of enzyme corresponds to
1 nmole of total nucleotides incorporated in one hour at 37°C.
Each value is the mean of two distinct determinations.   Control
cells: open bars;   irradiated cells: shaded bars.
The survival value of irradiated cells was 66% of the control
cells.

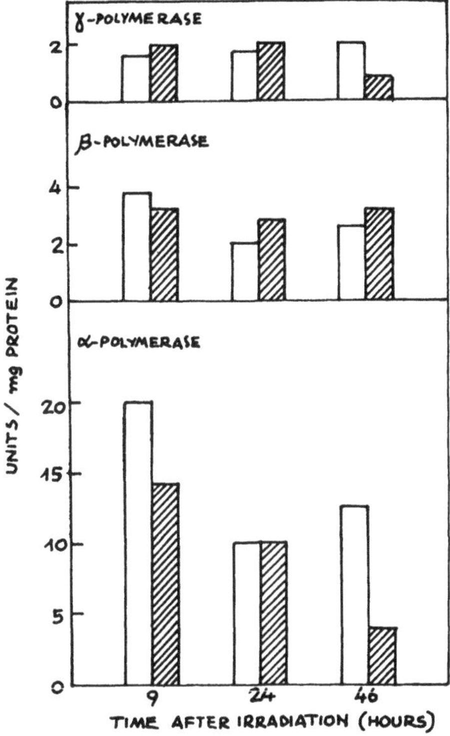

Figure 2. Levels of DNA polymerases α, β and γ in EUE cells UV-irradiated with 16 J/m^2.
For the determination of DNA polymerases activities see legend of Fig. 1. Control cells: open bars; irradiated cells: shaded bars. The survival value of irradiated cells was 25% of the control cells.

In order to check whether UV irradiation could have an inducing effect on a DNA polymerase we have determined the levels of α, β and γ polymerases at different times after the exposure in human heteroploid cells EUE and in human lymphocytes. In the experiment reported in Fig. 1 the EUE cells were irradiated with 6 J/m^2 and the levels of the three DNA polymerases determined after 4, 8, 24, 48 hours in irradiated and control cells, respectively. It is evident that no significant variation of the three polymerases is observed. At a higher dose of irradiation (16 J/m^2) a rather dramatic change is found for α-polymerase which decreases to very low levels at 48 hours after irradiation, whereas β and γ-polymerases tend to remain

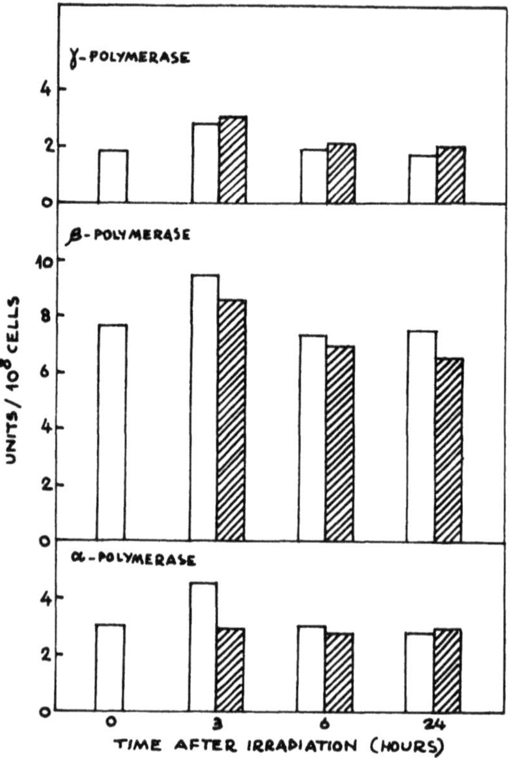

Figure 3. Levels of DNA polymerases α, β and γ in unstimulated human lymphocytes.
Lymphocytes, just separated from peripheral blood, were irradiated with 10 J/m^2 and pooled at 3, 6, 24 hours after irradiation. Non irradiated cells samples were collected at the indicated times and used as controls. For the determination of DNA
polymerase activities see legend of Fig. 1. Control cells: open
bars; irradiated cells: shaded bars.

constant (Fig. 2). This is consistent with the low cell survival
value (25% versus 66% of 6 J/m^2) obtained in this case, indicating that most of the cells stop dividing and hence loose α-polymerase. The results obtained with human unstimulated lymphocytes are reported in Fig. 3. It is evident that the amount of
α-polymerase is about 20 times lower than that found in dividing
cells (cf. Fig. 1 and 2) and that β-polymerase represents the
major activity. No significant variations were obtained after a

dose of 10 J/m^2, thus suggesting that UV-irradiation does not induce an increase in activity of one of the three polymerases, parallel to its effect on DNA repair.

Similar results have been obtained for the three DNA polymerases on UV-irradiated monkey kidney CV-1 confluent cells (59) whereas the DNA ligase activity increases several times (60). This observation is of particular interest since it shows that these two classes of enzymes (DNA polymerase α and larger form DNA ligase), which tend to increase in coordination during DNA synthesis, behave in a different way after UV-irradiation and subsequent repair.

A different approach has been recently followed successfully by Hübscher et al. for understanding the function of DNA polymerase in repair. They have isolated neuronal rat nuclei from mature animals and found that UV-irradiation brings about a 7-10 fold stimulation of DNA repair synthesis in this system (44). Since β-polymerase is virtually the exclusive DNA polymerase present in the nuclei at this developmental stage, they conclude that this enzyme is responsible for repairing UV-damaged nuclear DNA. On the contrary, irradiation and carcinogen treatment of synaptosomal mitochondria does not result in stimulation of DNA repair, suggesting either that excising enzymes are not present in these organelles (61) or that DNA polymerase γ may not be able to function as a repair enzyme.

The higher affinity of γ-polymerase for the deoxynucleotides (dNTPs) with respect to α and β-polymerases suggested a possible involvement of this enzyme in DNA repair since this process is rather insensitive to the lowering of the intracellular pool of dNTPs. However, this correlation might not be significant considering that (i) γ-enzyme is found mainly or solely in mitochondria, (ii) other factors could alter in vivo the affinity measured in vitro of the polymerase for the dNTPs and (iii) the nucleotide pool level of actively synthesizing cells is about 10 times lower than the Km value found for the γ-polymerase (62).

A summary of the possible correlations of the three DNA polymerases with DNA replication and repair is presented in Table II.

Table II

Correlation of DNA polymerases α , β and γ with
DNA replication and repair processes.

System	DNA replication			References
	α	β	γ	
Cell cultures and organs	+	–	–	2, 4, 6, 8[*]
SV40 DNA replication	+	-	-	11, 33, 34
ddTTP-treated HeLa cells	+	-	-	11, 35
Aphidicolin-treated sea urchins	+	-	-	36
Mammalian, avian mitochondria	-	-	+	16-19
Rat synaptosomal mitochondria	-	-	+	44
	DNA repair			
	α	β	γ	
PHA stimulated lymphocytes	-	+	-	45
Mature spermatocytes	-	+	-	48
Blood granulocytes	-	+	-	49, 50
Neuronal rat nuclei	-	+	-	44
ddTTP-treated rat liver	-	+		51
Bleomycin-induced breaks in DNA	-	+		53

[*] reviews

DNA polymerase and Fidelity
(see also L. Loeb, this volume)

An additional property characterizing the three mammalian DNA polymerases and distinguishing them from the prokaryotic enzymes, is the absence of associated DNases, that is of those exonucleolytic activities capable of recognizing and excising the nucleotides erroneously inserted (3' to 5' exo) and of excising pyrimidine dimers and other damaged DNA fragments (5' to 3' exo) (2). However, the fidelity of copying the template DNA by the DNA polymerase α and β is very high since in an _in vitro_ system no more than one error is made per 10,000 polymerized nucleotides (63, 64). The observation that the mammalian polymerases can use X-irradiated DNA as primer (65) and, unlike the bacterial enzymes, are able to copy DNA across pyrimidine dimers on _in vitro_ templates (63), would make one expect a high number of mutations induced by X and UV irradiations. Since this phenomenon does not occur at the cellular level, where even a strong inhibition of DNA replication is observed, one has to conclude that the DNA polymerases might interact _in vivo_ with other enzymes and factors to form a more complex system having the capacity to recognize DNA alterations and to correct them in a specific way. Experimental data supporting this hypothesis has been recently presented by Radman et al. (66) who purified the 3' to 5' exonuclease from calf spleen and showed that, when this enzyme is added to DNA polymerase α , it both stimulates the polymerase activity on a terminally mismatched template and induces an arrest of DNA synthesis on a UV irradiated template, thus mimicking _in vitro_ the effect usually obtained with the _E. coli_ DNA polymerase I. This suggests that even the mammalian replication machinery is unable to copy pyrimidine dimers _in vivo_.

CONCLUSIONS

It is now becoming clear that the functions of α- and β-polymerases are centered on the replication and repair of nuclear DNA respectively, while γ-polymerase performs replication of mitochondrial DNA. However, some words of caution are needed: at the present time it cannot be excluded that α-polymerase also part cipates in the process of DNA repair and β-polymerase could be necessary in replication by filling gaps after the removal of primer RNA's; γ-polymerase is found in small amounts also in the nucleus, though this must be ascertained by more precise techniques. Finally the possibility still exists that the DNA polymerase involved in DNA repair is different from the known α, β and γ-polymerases.

ACKNOWLEDGEMENTS

We wish to thank A. Falaschi and F. Nuzzo for critical readings of the paper and S. Spadari for providing manuscripts before publication. This work was partially supported by EURATOM (Contract 125-74-I-BIOI). This publication is contribution n° 1587 of the Biology, Radiation Protection and Medical Research Division of European Communities. A. I. Scovassi is presently at the Laboratoire de Biochimie, Fondation Curie, Paris.

REFERENCES

(1)   Weissbach, A., Baltimore, D., Bollum, F. J., Gallo, R. C. and Korn, D. (1975) Science 190:401-402

(2)   Bollum, F. J. (1975) Prog. Nucl. Acid Res. and Mol. Biol. 15:109-144

(3)   Holmes, A. M. and Johnston, I. R. (1975) FEBS Letters 60:233-243

(4)   Weissbach, A. (1977) Ann. Rev. Biochem. 46:25-47

(5)   Wintersberger, E. (1977) TIBS 58-61

(6)   Falaschi, A. and Spadari, S. (1978) from: DNA SYNTHESIS: Present and Future, Edited by I. Molineux and M. Kohiyama (Plenum Publishing Corporation, 1978)

(7)   Dube, D. K., Travaglini, E. C. and Loeb, L. A. (1978) Methods in Cell Biology XIX. Ed. by G. Stein, J. Stein, L. J. Kleinsmith

(8) Sarngadharan, M.G., Robert-Guroff, M. and Gallo, R.C.
     (1978) Biochim. Biophys. Acta 516:419-487

(9) Bollum, F.J. (1979) from: Antiviral Mechanisms for the
     control of neoplasia, ed. by E.S.H. Prakash, Plenum
     Publishing Co. (1979)

(10) Ohashi, M., Taguchi, T. and Ikegami, S. (1978)
     Biochem. Biophys. Res. Comm. 82:1084-1090

(11) Edenberg, H.J., Anderson, S. and De Pamphilis, M.L.
     (1978) J. Biol. Chem. 253:3273-3280

(12) Chang, L.M.S. (1973) J. Biol. Chem. 248:3789-3795

(13) Wang, T.S-F., Sedwick, W.D. and Korn, D. (1975)
     J. Biol. Chem. 250:7040-7044

(14) Chang, L.M.S. (1976) Science 191:1183-1185

(15) Knopf, K.W., Yamada, M. and Weissbach, A. (1976)
     Biochemistry 15:4540-4548

(16) Bolden, A., Pedrali-Noy, G. and Weissbach, A. (1977)
     J. Biol. Chem. 252:3351-3356

(17) Bertazzoni, U., Scovassi, A.I. and Brun, G. (1977)
     Eur. J. Biochem. 81:237-248

(18) Hübscher, U., Kuenzle, C.C. and Spadari, S. (1977)
     Eur. J. Biochem. 81:249-258

(19) Tarrago-Litvak, L., Viratelle, O., Darriet, D.,
     Dalibart, R., Graves, P.V. and Litvak, S. (1978)
     Nucleic Acid Res. 5:2197-2210

(20) Bollum, F.J. and Potter, V.R. (1957) JACS 79:3603

(21) Loeb, L.A., Agarwal, S.S. and Woodside, A.M. (1968)
     Proc. Natl. Acad. Sci. USA 61:827-834

(22) Rabinowitz, Y., McCluskey, I.S., Wong, P. and Wilhite,
     B.A. (1969) Exp. Cell Res. 57:257-262

(23)   Pedrini, A.M., Nuzzo, F., Ciarrocchi, G., Dalprà, L.
        and Falaschi, A. (1972) Biochem. Biophys. Res.
        Commun. 47:1221-1227

(24)   Tyrsted, G., Munch-Petersen, B. and Cloos, L. (1973)
        Exp. Cell Res. 77:415-427

(25)   Chang, L.M.S. and Bollum, F.J. (1972) J.Biol.Chem.
        247:7948-7950

(26)   Chang, L.M.S., Brown, Mc.K. and Bollum, F.J.
        (1973) J.Mol.Biol. 74:1-8

(27)   Coleman, M.S., Hutton, I.J. and Bollum, F.J. (1974)
        Nature 248:407-409

(28)   Roodman, G.D., Hutton, J.J. and Bollum, F.J. (1976)
        Biochim.Biophys.Acta 425:478-491

(29)   Spadari, S., Villani, G. and Hardt, N. (1978) Exp.
        Cell Res. 113:57-62

(30)   Spadari, S. and Weissbach, A. (1974) J.Mol.Biol.
        86:11-20

(31)   Chiu, R.W. and Baril, E.F. (1975) J.Biol.Chem.
        250:7951-7957

(32)   Hübscher, U., Kuenzle, C.C. and Spadari, S. (1977)
        Nucleic Acids Res. 4:2917-2929

(33)   Otto, B. and Fanning, E. (1978) Nucleic Acids Res.
        5:1715-1728

(34)   Mechali, M., Girard, M. and de Recondo, A.M. (1977)
        J.Virology 23:117-125

(35)   Waqar, M.A., Evans, M.J. and Huberman, J.A. (1978)
        Nucleic Acids Res. 5:1933-1946

(36)   Ikegami, S., Taguchi, T., Ohashi, M., Oguro, M., Naga-
        no, H. and Mano, Y. (1978) Nature 275:458-460

(37) Spadari, S. and Weissbach, A. (1975) Proc. Natl. Acad. Sci. USA 72:503-507

(38) Edenberg, H. J. and Huberman, J. A. (1975) Ann. Rev. Genetics 9:245-284

(39) Eliasson, R. and Reichard, P. (1978) Nature 272:184-185

(40) Brun, G. and Weissbach, A. (1978) Proc. Natl. Acad. Sci. USA 75:5931-5935

(41) Herrick, G., Delius, A. and Alberts, B. (1976) J. Biol. Chem. 251:2142-2146

(42) Otto, B., Baynes, M. and Knippers, R. (1977) Eur. J. Biochem. 73:17-24

(43) Cobianchi, F., Riva, S., Mastromei, G., Spadari, S., Pedrali, G. and Falaschi, A. (1978) Cold Spring Harbor Symp. Quant. Biol., vol. 43 "DNA: Replication and Recombination", in press

(44) Hübscher, U., Kuenzle, C. C. and Spadari, S. (1979) Proc. Natl. Acad. Sci. USA, in press

(45) Bertazzoni, U., Stefanini, M., Pedrali-Noy, G., Giulotto, E., Nuzzo, F., Falaschi, A. and Spadari, S. (1976) Proc. Natl. Acad. Sci. USA 73:785-789

(46) Pedrali-Noy, G., Dalprà, L., Pedrini, A. M., Ciarrocchi, G., Giulotto, E., Nuzzo, F. and Falaschi, A. (1974) Nucleic Acids Res. 1:1183-1200

(47) Dalprà, L., Stefanini, M., Giulotto, E., Falaschi, and Nuzzo, F. (1978) Haematologica 64:31-39

(48) Hecht, N. B., Farrell, D. and Davidson, D. (1976) Dev. Biol. 48:56-66

(49) Ringborg, U. and Lambert, B. (1977) Cancer Letters 3:77-81

(50)   Coleman, M. S., Hutton, J. J. and Bollum, F. J. (1974)
        Blood 44:19-32

(51)   Smith, G. J., Charlton, R. K., Grisham, J. W. and
        Kaufman, D. (1978) Biochem. Biophys. Res. Comm.
        4:1538-1544

(52)   Fisher, P. A., Wang, T. S. F. and Korn, D. (1978) Cold
        Spring Harbor Symp. Quant. Biol. vol. 43 "DNA Replica-
        tion and Recombination" in press

(53)   Coetzee, M. L., Chou, R. and Ove, P. (1978) Cancer
        Res. 38:3621-3627

(54)   Setlow, R. B. (1978) Nature 271:713-717

(55)   Arlett, C. and Lehmann, A. (1978) Ann.Rev.Genet. 12:95-115

(56)   Bertazzoni, U., Stefanini, M., Pedrali-Noy, G., Nuzzo,
        F., and Falaschi, A. (1977) Nucleic Acids Res.
        4:141-148

(57)   Bertazzoni, U., Scovassi, A. I., Stefanini, M., Giulotto,
        E., Spadari, S. and Pedrini, M. A. (1978) Nucleic Acids
        Res. 5:2189-2196

(58)   Parker, V. P. and Lieberman, M. W. (1977) Nucleic
        Acids Res. 4:2029-2037

(59)   Wicker, R., Scovassi, A. I. and Nocentini, S. (1978)
        submitted for publication

(60)   Mezzina, M. and Nocentini, S. (1978) Nucleic Acids Res.
        5:4317-4328

(61)   Clayton, D. A., Doda, J. M. and Friedberg, E. C. (1974)
        Proc. Natl. Acad. Sci. USA 71:2777-2781

(62)   Bray, G. and Brent, T. P. (1972) Biochim. Biophys. Acta
        269:184-191

(63) Villani, G., Defais, M., Spadari, S., Caillet-Fauquet,
      P., Boiteux, S. and Radman, M. (1977) from "Research
      in Photobiology" ed. by A. Castellani, Plenum Publishing
      Co., New York

(64) Loeb, L. (1979) this volume

(65) Campagnari, F., Bertazzoni, U. and Clerici, L. (1967)
      J. Biol. Chem. 242:2168-2171

(66) Radman, M., Villani, G., Boiteux, S., Kinsella, A.R.,
      Glickman, B.W. and Spadari, S. (1978) Cold Spring
      Harbor Symp. Quant. Biol., vol. 43 "DNA: Replication
      and Recombination", in press

DNA SYNTHESIS AND DNA POLYMERASES IN TONSILLAR LYMPHOCYTES

F. Antoni and Maria Staub

1st Institute of Biochemistry

Semmelweis University Medical School, Budapest, Hungary

INTRODUCTION

The genetic message represents the essence of a living system, containing the information necessary for the perpetuation of the species. In eukaryotic cells the genetic message that genes are composed of is the DNA. There is very clear evidence that RNA may also serve as a genetic message in viruses, and it cannot be excluded that certain RNA molecules may have the same role in eukaryotic cells. Genetic information can be transmitted in two ways. One of these is the transfer of DNA molecules from one cell to another: via free DNA (transformation), via viral vector (transduction), or by direct cell contact (conjugation). The three above procedures have been very extensively studied in the haploid prokaryotes. There are reports of successful transfer of bacterial genes to plant cells and cell fusion. The cell fusion may be considered as a form of message transmission. The second way for the transmission of genetic information is the replication of the genetic material, i.e. the process by which one message will be converted into two identical informations. The process of the replication of the genetic material has attracted considerable interest over the past years. Both, prokaryotic and eukaryotic DNA appear to be replicated in essentially the same semiconservative way. Many details of the DNA replication are well known, however, less data are available about how the process is regulated.

The preservation of the ability of proliferation is probably the most characteristic feature of the cell population responsible for the immune response. Replication of cellular DNA is induced in vivo by various immunological stimuli, however, it can also be initiated in vitro by the stimulation of cultured cells by nonspecific

mitogenic substances. Thus the replication of cellular DNA can be
investigated in well-defined model systems.

The relationships between DNA synthesis, and DNA polymerases
are dealt with in the present study. DNA synthesis of eukaryotic
cells has been investigated by many workers, nevertheless, there
are several unresolved problems. Deoxyribonucleosides from the
nutrient medium and endogeneous precursors are phosphorylated to
triphosphates at the expense of ATP by specific kinase enzymes.
Deoxyribonucleoside triphosphates (dNTP) are polymerized to new
DNA chains by the DNA polymerase enzymes as directed by the template DNA molecule. However, the specific biochemical signal initiating DNA synthesis in some part of the chromatin is not known,
and the enzyme responsible for the initiation of the template-de-
pendent polymerization process has not been identified either
(Weissbach, H. 1977, Reichardt, P. 1977). Reproduction of the genetic material is one of the most important and intricate biological phenomena, and perhaps this is why several DNA-polymerizing
enzymes exist. The term DNA polymerase (DNA nucleotidyl transferase E.C. 2.7.7.) refers to an enzyme capable of the template-de-
pendent polymerization of dNTPs. The DNA-dependent DNA polymerases
and the RNA-dependent DNA polymerase differ as regards the template
required. Three types of DNA-dependent DNA polymerases have been
demonstrated up to now in mammalian cells. Distinguished by their
intracellular localization, molecular weight and enzymatic properties these are the cytoplasmic (α), nuclear (β) and mitochondrial
(mt) DNA polymerases (Holmes and Johnston 1975, Weissbach, A.1977).
The cytoplasmic enzyme is supposed to consist of several subunits;
its molecular weight is between 1 to $3\text{-}10^5$ daltons, and its activity is inhibited by Sh-blocking agents. The nuclear DNA (β) polymerase is not sensitive to SH reagents; its molecular weight is
$3\text{-}5\text{x}10^4$ daltons, and it consists of a single polypeptide chain. It
is firmly bound to the chromatin, and can be solubilized only with
1 M of KCl. The molecular weight of the mitochondrial enzyme is
similar to that of the cytoplasmic DNA polymerase.

The reaction of RNA-dependent DNA polymerization has been discovered in connection with the oncogenic RNA viruses. The enzyme
promoting this reaction has been termed reverse transcriptase. The
appearance of the reverse transcriptase is due to infection by RNA
viruses. However, the presence of RNA-dependent DNA polymerases has
also been demonstrated in apparently intact HeLa cells (Fridlender
et al. 1972), lymphoblasts (Lewis et al. 1974a, b), mouse myeloma
cells (Matsukage et al. 1974), and in several other types of cells
and tissues (Weissbach, 1977). About 1% of the total DNA polymerase
activity is due to the RNA-dependent enzyme in eukaryotic cells
(Spadari and Weissbach, 1974a, b, Weissbach, 1977). Moreover, antibodies raised against the viral reverse transcriptase do not inac-
tivate the RNA-dependent DNA polymerase isolated from intact cells,

therefore the latter has to be regarded as a different enzyme
called R-DNA polymerase or, more recently, γ-DNA polymerase. No
immunological cross reaction has been found with α, β, and γ-enzy-
mes, either. Recently, DNA polymerase activity of mitochondria has
been found to be closely related, or identical to the DNA polymerase γ activity (Bolden et al. 1977).

The properties and the nomenclature of the eukaryotic DNA
polymerases are summarised in Table I.

Table I.   Nomenclature for the eukaryotic cell DNA polymerases
           by A.Weissbach, Ann.Rev.Biochem. 1977. 46, 25-47.

DNA polymerase	Molecular weight	Inhibition by N-ethyl maleimide	Salt effect
alfa (α)	120.000-300.000	+	Inhibited at NaCl concentration above 25 mM
beta (β)	30.000-50.000	-	Stimulated by 100-200 mM NaCl but inhibited by 50 mM $PO_4 3^-$
gamma (γ)	150.000-300.000	+	Stimulated by 100-250 mM KCl and 50 mM $PO_4 3^-$
Mitochondrial	150.000	+	Stimulated by 100-200 mM KCl

The DNA polymerases described above require for their activity
the four dNTPs as substrate, the presence of magnesium ions, template and a primary DNA segment to which the subsequent units are
attached as directed by the template. Another enzyme called terminal transferase) is also regarded as a DNA polymerase, however,
this enzyme performs the elongation of DNA chains (at the 3'-hy-
droxy terminal) without or independent of any template. This enzyme
was first demonstrated in calf thymus by Chang and Bollum (1971).
Apparently, the terminal transferase occurs exclusively in lymphoid
tissues. It has been detected in peripheral cells in lymphoblastic
leukaemia (McCaffrey et al. 1973), chronic myeologenous leukaemia
(Coleman et al. 1974) and in the case of acute myelomonocytic leukaemia (Sarin and Gallo, 1974), and in low amount in the bone marrow. The presence of the enzyme in undifferentiated lymphoid cells
and in the thymus has suggested that it may have some specific role
in the development of the immune response. Baltimore (1974) has
supposed that the template-independent terminal transferase might
be responsible for the generation of "mutations" in the genes cod-
ing for the variable segments of immunoglobulins. The activity of
the enzyme, thereby the variations, might be induced by the antigen

itself through some unknown mechanism. Although the actual role of
the terminal transferase is not yet known certainly, the circumstance that it does occur only in proliferating, undifferentiated
lymphoid tissue indicates the importance of this reaction in immunological events.

Study of the mechanism of DNA replication has largely been
promoted by the isolation of mutant bacteria defective in one or
another DNA polymerase enzyme. However, this approach has not been
successful with eukaryotic cells, perhaps because of the greater
complexity of DNA replication in higher organisms. Another approach
to the problem is offered by the investigation of resting and proliferating eukaryotic cells with respect to the enzymes of DNA
synthesis. It has been demonstrated that the amount of the relevant
enzymes changes with the phases of the mitotic cycle: the thymidine
kinase and the cytoplasmic DNA polymerase exhibit increased activity already before the period of DNA synthesis (S phase) (Smith and
Gallo, 1972, Spadari and Weissbach, 1974a, Craig et al. 1975,
Tyrsted and Munch-Petersen, 1977). Non-malignant, normal, differentiated cells do not proliferate in vitro but go into resting state.
However, the proliferation of lymphocytes can be induced by mitogens in vitro, thus the synthesis of DNA can be investigated in
normal cultured cells.

Human palatine tonsils represent a convenient source of lymphocytes; 10^9 viable cells can be obtained from one pair of tonsils
offering enough material of human origin for biochemical experiments on a preparative scale (Piffkó et al. 1970, Antoni and Staub,
1978).

The cell population isolated from the tonsil contains 95-97%
lymphoid cells and only 2-5% phagocytic cells (Greaves et al. 1974).
However, tonsillar lymphocytes are far from being uniform, since
the cells are in various stages of proliferation (Merler and
Silberschmidt, 1972, Geha and Merler, 1974, Gatien et al. 1975,
Staub et al. 1975).

RESULTS

ISOLATION AND CHARACTERIZATION OF TONSILLAR LYMPHOCYTES EXHIBITING
SPONTANEOUS DNA SYNTHESIS

Heterogeneity of the freshly isolated tonsillar lymphocyte
population has suggested that it also contains S phase cells. Whereas only the mitotic phase has its morphological characteristics, S
phase cells are distinguished only by their ability to incorporate
^{3}H-thymidine into DNA.

The time course of thymidine incorporation was investigated
by both continuous (Fig.1a) and pulse (Fig.1b) labelling techniques.

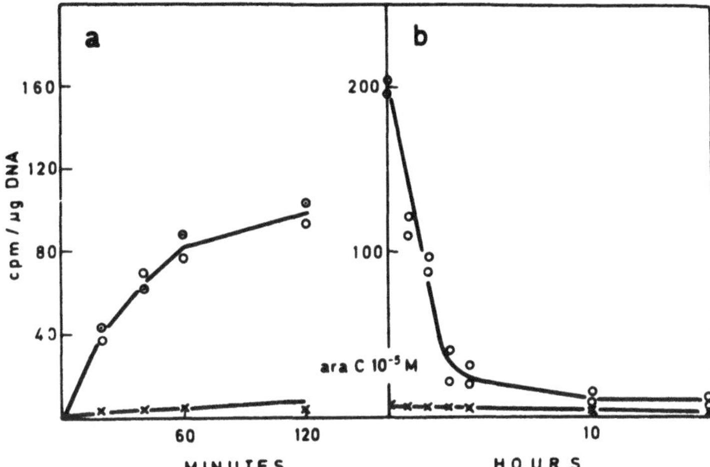

Fig. 1. Incorporation of ^{3}H-thymidine into tonsillar lymphocytes in short-term and long-term cultures. Inhibition of incorporation by arabinosul-cytosine (ara-C). Tonsillar cells (2×10^6/ml) were incubated in Eagle's MEM in the presence of ^{3}H-thymidine (1.5 µCi/ml; sp.a. 25 Ci/mmole), and ara-C (10^{-5}M). (a) Short-term culture, continuous labelling, (b) Long-term culture, pulse labelling for periods of 20 min. Additions: (O) none; (x) ara-C.

Parallel samples containing arabinosyl-cytosine (ara-C) were run in order to exclude any interference by bacterial contaminants. Ara-C is known to inhibit the DNA synthesis of eukaryotes only, while it does not affect bacterial DNA synthesis (Staub et al. 1975, Magnusson et al. 1974). In our experiments the incorporation of ^{3}H-thymidine was completely blocked by ara-C, proving that the incorporation was due to the DNA synthesis of lymphocytes, i.e. microbial contamination was effectively prevented by careful preparation (Staub et al. 1975). The rate of incorporation of ^{3}H-thymidine was found to decrease with time and it came to a standstill in about 5 hours.

There are several methods for the separation of different subpopulations of lymphocytes, most of which exploit differences in the surface characteristics of the cells, while others are based on the differences in the sedimentation velocity of lymphocytes. The latter involve the centrifugation of cells in a density gradient prepared from solutions of various polymeric substances. Serum albumin has been found to be a suitable solute for this purpose

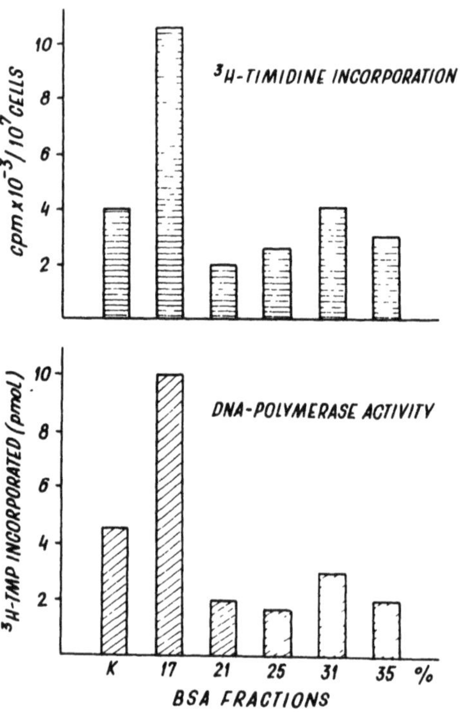

Fig. 2. Incorporation of ^{3}H-thymidine, DNA polymerase activity and cAMP content of tonsillar cell fractions obtained by density gradient centrifugation. Labelling was performed as described in Fig.1; assay of DNA polymerase is described in Γig.11. Estimation of cAMP is described by Faragõ et al. (1974).

(August et al. 1970, Gatien et al. 1975). We adopted the method of August et al. with several modifications. Tonsil lymphocytes (5x10^8) were suspended in Hanks' solution containing 8% bovine serum albumin (BSA) and layered onto a density gradient consisting of 35, 31, 25, 21 and 17% BSA solutions. After centrifugations, the cells were recovered from the interfaces of the gradient.

The cells recovered from the different fractions were assayed for their ability to incorporate ^{3}H-thymidine, DNA polymerase activity and cAMP content (Fig.2). Lymphocytes exhibiting high DNA synthetic activity were found in the "light", slowly sedimenting fractions, however, none of the fractions was inactive in this respect. The separation of cells was not complete, but this could not be expected in view of the wide variation in the size of proliferating cells (7-22 μm in diameter) being in various stages of the cell cycle. The close correlation between ^{3}H-thymidine incorporation and DNA polymerase activity indicated the presence of S phase cells (Staub, 1974).

Tonsillar cells were separated into nine fractions by Merler and co-workers (Merler and Silberschmidt, 1972, Gatien et al. 1975). Cells exhibiting spontaneous DNA synthesis were recovered also in the light fractions. These cells were shown to carry surface immunoglobulins, and they were not stimulated by PHA, and did not form E rosettes. The cells described were considered by Merler and his co-workers to be "precursor" cells in an early stage of differentiation.

The amounts of cyclic AMP and cGMP change in opposite direction during the cell cycle in the synchronized HeLa cells. (Seifert and Rudland, 1974). As shown in Fig.2, the cAMP content of the cells in the fractions of tonsillar lymphocytes correlated with the DNA synthetic activity of the cells (Staub et al. 1975). The relationship between the synthesis of cAMP and DNA is not known.

Tonsil lymphocyte subpopulations were also separated according to their differences in the surface characteristics. Separation on nylon-fiber columns was basically carried out according to Greaves and Brown (1974). The immunological characterisation of the adherent and non-adherent cell fraction is shown in Fig.3. As can be seen, the non-adherent cells consisted mainly of T lymphocytes, while the mechanically removed adherent fraction contained only 7% E-rosette forming cells.

The ^{3}H-thymidine uptake and incorporation and also the DNA polymerase activity was 3-5 times higher in the adherent cell fraction, as in the non-adherent one (Fig.4). Thus, in tonsils the B-cell enriched fraction turned out to be more active in DNA synthesis, than the T cell enriched population.

CHANGES OF THE TONSILLAR LYMPHOCYTE POPULATION IN CULTURE, EXCRE-
TION OF DNA

Most of the investigations of lymphocyte proliferation and blast transformation have been performed on cells stimulated in vitro. At variance with these examinations, the proliferation of tonsillar cells was apparently induced in vivo already. Little is known about the result of the proliferation process and of the fate of the cells removed from their natural environment before the

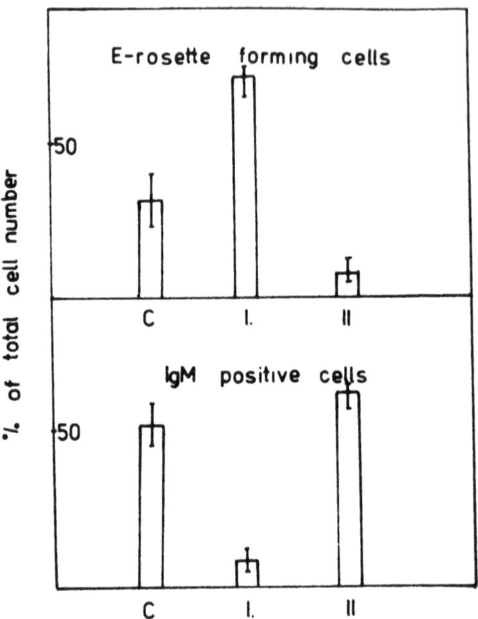

Fig. 3. Immunological characterization of nylon-wool separated cells. Separation on nylon-fiber column the estimation of Surface-IgM bearing lymphocytes and the quantitation of T-lymphocytes was carried out as described earlier (Staub et al. 1978). C: unseparated, control cells. I: non-adherent cells ("T"). II: adherent cells ("B").

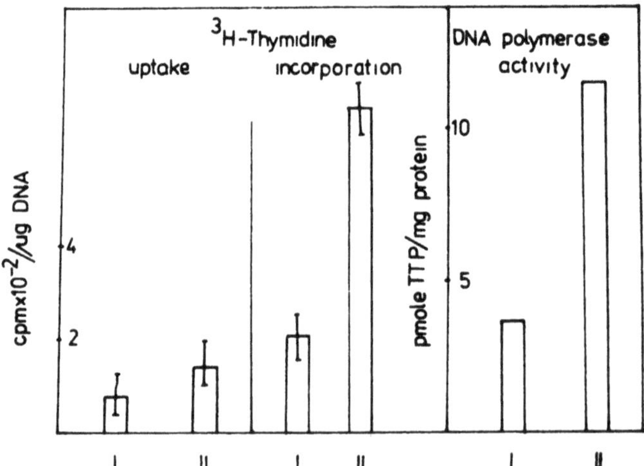

Fig. 4. The ^{3}H-thymidine uptake and incorporation, and the DNA polymerase activity of nylon-wool separated lymphocytes. I: non-adherent cells ("T"). II: adherent cells ("B").

completion of mitosis. In order to answer these questions, we examined the changes in the morphological and biochemical characteristics of cultured tonsillar lymphocytes as compared to freshly isolated cells.

Average diameter of freshly isolated tonsillar lymphocytes was 13 μm. This value was reduced to 6 μm after 20 h in culture. The cell size decreased in close correlation with the ^{3}H-thymidine uptake of the cells as shown by the labelling index determined by autoradiography (Fig.5). In the presence of phytohaemagglutinin (PHA) the decrease of cell diameter was followed by an increase of this value to 10 μm at 80 h; also the number of labelled cells increased.

Considering individual variations, freshly isolated tonsillar lymphocytes contained 15-30% blast cells, 50-70% middle-sized and large lymphocytes and 10-35% small lymphocytes. The number of large-sized cells decreased in the culture even in the presence of PHA. About 95% of the cells transformed into small lymphocytes after 20 h in culture. Reappearance of blast cells was observed in the

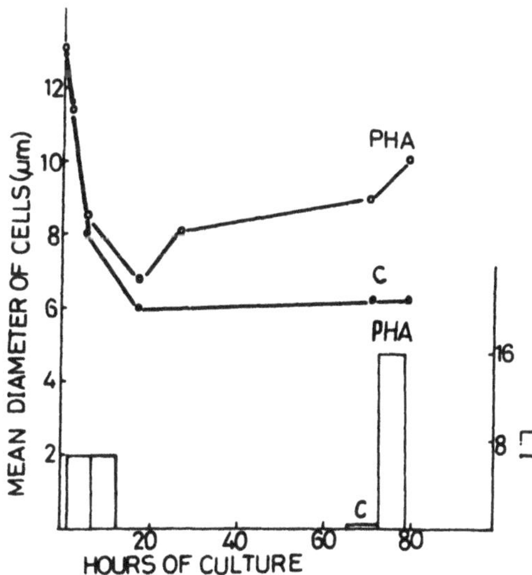

Fig. 5. Changes in the average cell diameter and labelling index of cultured tonsillar lymphocytes. Averages of the diameters of 200 cells are shown (a). Labelling was performed by ^{3}H-thymidine (0.5 μCi/ml; sp.a. 5 Ci/mmole, for 6 hours. Labelling index was determined by autoradiography (b).

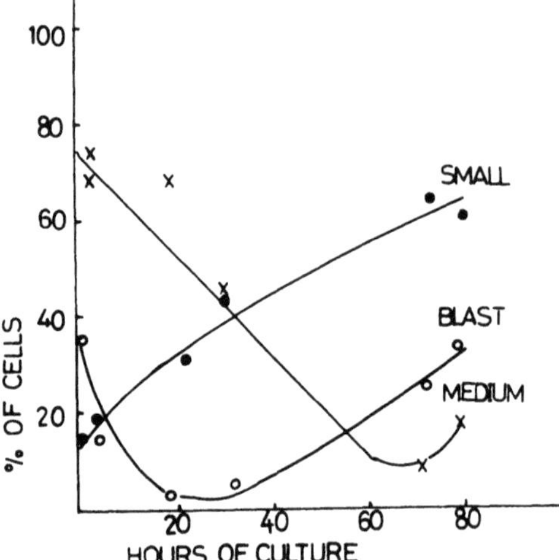

Fig. 6. Changes in cell types in the culture of tonsil lymphocytes.
Experimental conditions were as described in Staub et al. (1976).

PHA-containing cultures on the third day (Fig.6). The viability of
the cultured cells was investigated by the trypane blue exclusion
test. After three days in culture 80-90% of the cells excluded the
dye indicating that no significant decrease of viability occurred
under the conditions applied.

It has been concluded that the heterogeneous population of
freshly isolated tonsillar lymphocytes transforms spontaneously into a homogeneous small lymphocyte population in culture (Staub et
al. 1976).

The release of DNA into the culture medium has been observed
accompanying PHA-stimulation of blood lymphocytes (Rogers et al.
1972, 1976; Anker et al. 1976; Hoesli et al. 1977). The most striking difference between tonsil and peripheral lymphocytes is a high
"spontaneous" DNA synthesis in the former. Therefore, tonsil lymphocytes offer a good possibility to decide whether non-specific
stimulation is a prerequisite of DNA release with lymphocytes; or
has been the transformation into resting lymphocytes accompanied
by loss in DNA content too?

When human tonsil lymphocytes were cultured over a three-day
period the number of viable cells remained within ± 20% of the initial count. At least 90% of the cells counted on any day for three
days, were viable as judged by trypane blue exclusion. However, the
amount of DNA in the cells, as measured by the diphenylamine reaction (Burton, 1968), either dropped after three days, or remained
unchanged.

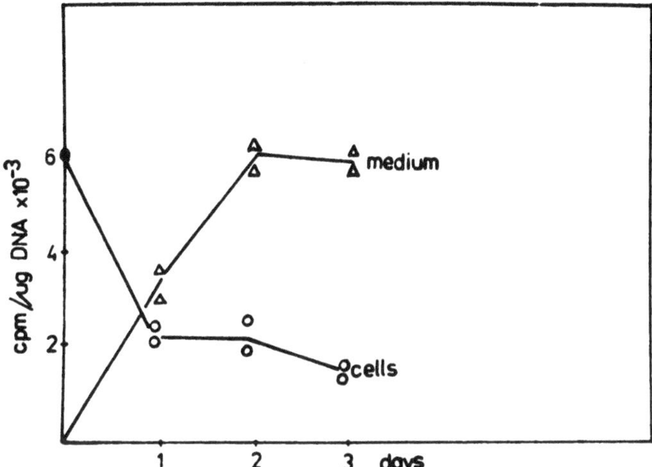

Fig. 7. Changes in the specific activity of acid-precipitable ^{3}H-thymidine content of freshly isolated cells (o-o-o) and their medium (-▲-▲). (Staub and Antoni, 1978b).

The amount of newly synthesized DNA in the cells decreased over the culture period. This decrease of newly synthesized cellular DNA was accompanied by a concomitant increase in acid-precipitable counts in the medium (Fig.7), in amounting on the 3rd day of culture to about 80 to 90% of the intracellular counts measured at the beginning of the experiment. PHA apparently did not influence the loss of newly synthesized DNA from the cells.

Attempts to determine cell counts and percent of cells excluding trypane blue at different times, proved to be non-reproducible, neither specific nor sensitive enough. Therefore, the selective nature of DNA loss was considered as an indication of cell viability. Experiments were carried out in which non-stimulated lymphocytes were pulsed with ^{3}H-thymidine, ^{3}H-uridine and ^{14}C-valine in parallel cultures. The results of a representative experiment are shown in Fig.8. The cells were pulsed immediately after isolation and then the fate of the three isotopes was followed for 3 days. About 25% of the acid-precipitable ^{3}H-thymidine was lost from the cells during the first 24 hours after the pulse label. In contrast during the same 24 h period, the amount of intracellular acid-precipitable ^{3}H-uridine increased in each instance, and the ^{14}C-valine did not change or also increased to a lesser extent.

The increase observed in acid-precipitable ^{3}H-uridine during the first 24 h may be explained by assuming the existence of a large intracellular pool of acid-soluble ribonucleotides in human lymphocytes (Staub et al. 1978), which was converted by the cells

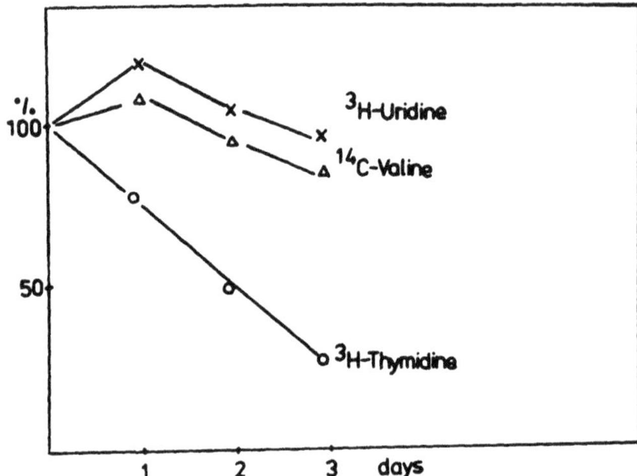

Fig. 8. Changes in the amounts of acid-precipitable ^{3}H-thymidine o-o-o, ^{3}H-uridine x-x-x, and ^{14}C-valine Δ - Δ - Δ, in non-stimulated tonsil lymphocytes during culture. The culture conditions and determinations were described by Staub and Antoni (1978b).

to an acid-precipitable form. A much lower pool of ^{3}H-thymidine has been found in lymphocytes and also in other cells (Trysted et al. 1977; Menth et al. 1976). Obviously, some lysis or desintegration of the cells must be taken into consideration (Bernheim et al. 1977), since the presence of cytotoxic lymphocyte populations cannot be excluded.

Although no clear explanation can be given at present for the release of DNA from lymphocytes stimulated either specifically in vivo or non-specifically in vitro, a number of possibilities exist. Rogers suggested (1976), that DNA excretion has a role in the immune response since it might reflect gen amplification. In this case, however, the high heterogeneity of DNA excreted by PHA-stimulated lymphocytes, remained unanswered.

Data published up till now suggest that the release of DNA is specifically characteristic for lymphocytes. However, recently DNA release from nuclei of rat hepatocytes has also been observed in the presence of heparin. This release was accompanied also by a disappearance of histones (Demidenko, 1977).

Concerning the excretion of DNA from lymphocytes the possibility of activation and release of a latent virus has also been suggested, especially endogenous C type viruses. These viruses are assumed to be involved in the immune response (Moroni and Schuman,

1977). Moreover, the release of DNA from lymphocytes has also been demonstrated in vivo after stimulation by bacterial lypopolysaccharides (Izui et al. 1977).

We feel that apart from all the above hypotheses there is a much simpler explanation for a release of DNA from lymphocytes. It might also be supposed that excretion of DNA by lymphocytes is a phenomenon by which the cells get rid of their "extra DNA" by reverting from a proliferating state to their resting state in culture. Further experiments are necessary to test the validity of this hypothesis.

THE ACTIVITY OF DNA POLYMERASES IN NON-STIMULATED AND PHA-STIMULATED TONSILLAR LYMPHOCYTES

As regards the cause of the cessation of DNA synthesis of cultures, we have no clear explanation. There were, however, two possible causes to be tested. The decrease of DNA synthesis may have been due to the inactivation of the DNA-polymerizing enzymes, or the culture conditions were unsatisfactory to maintain the synthetic activity of the S phase lymphocytes.

We estimated the total DNA polymerase content of the tonsillar lymphocytes by assaying the enzyme activity of the 100.000 g supernatant of disrupted cells, which contained the nuclear and cytoplasmic enzymes of the lymphocytes. The specific DNA polymerase activity of tonsillar cells (as related to total cellular protein) was similar to that of the spleen and of the regenerating liver, which contain proliferating cells. However, the specific activity of leukaemic lymphoblasts was found to be four times higher than the value of tonsil lymphocytes. In turn, the specific DNA polymerase activity of tonsillar lymphocytes is four times higher than that of lymphocytes isolated from the peripheral blood. The ratios mentioned allowed the rough estimation of the amount of S phase cells present in the tonsillar cell population, since it has been reported that the DNA synthetic phase is preceded and accompanied by an increase in the activity of the polymerase enzymes (Matsukage et al. 1974; Holmes et al. 1974; Spadari and Weissbach, 1974a, b).

The DNA polymerase activity of freshly prepared non-stimulated and PHA-stimulated cultured cells is shown in Fig.9a. The polymerase activity of non-stimulated cells decreased to about 50% of the initial value in three days, however this change could not account for the rapid cessation of ^{3}H-thymidine incorporation (Fig.9b). The incorporation of thymidine increased in the PHA-stimulated cultures from the second day on, culminating on the third day. It was accompanied by an increase in DNA polymerase activity, too. In the experiment shown in Fig.9 PHA was present in the stimulated culture from the beginning of the incubation, however, similar stimulation resulted when PHA was added 24 h later, although at that time the

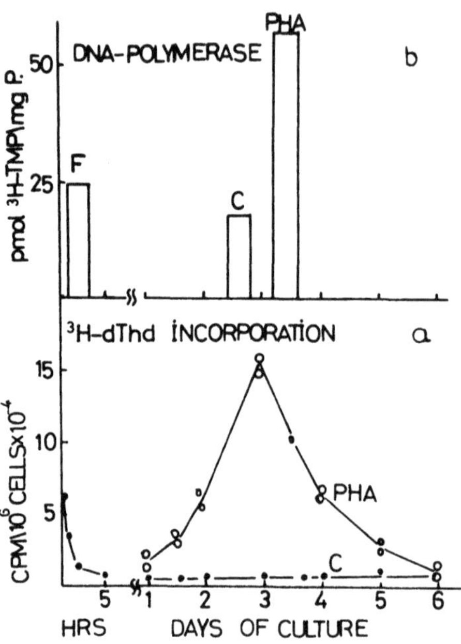

Fig. 9. DNA polymerase activity and ³H-thymidine incorporation of freshly isolated PHA-stimulated tonsillar lymphocytes. Tonsillar cells (2x10⁶/ml) were cultured in Eagle's MEM containing 10% human AB serum. DNA polymerase specific activity (a) or control (C) and PHA-stimulated cells (PHA) was determined in freshly prepared and in cultured (72 h) cells; enzyme activity was assayed as described in Fig.11. Incorporation of ³H-thymidine (b) by control and PHA-stimulated cells was measured by the pulse labelling technique as described in Fig.1.

culture contained small lymphocytes only. The latter finding was regarded as a proof for the viability of the cultured cells even when transformed back into small lymphocytes.

We wish to comment on the PHA-induced increase of the DNA poly-merase activity as follows. On the third day of culture the cell

suspensions contained only small lymphocytes, i.e. resting phase
cells only. Matsukage et al. (1974) and Spadari and Weissbach
(1974a, b) demonstrated by experiments on synchronized HeLa cells
that the amount of the nuclear polymerase is constant throughout
the cell cycle, and it is the cytoplasmic enzyme which increases
in amount during proliferation. Accordingly, in our experiment the
enzyme activity of non-stimulated cells reflected the amount of the
nuclear DNA polymerase on the third day of culture. The amount of
cytoplasmic enzyme in the freshly isolated cells was the difference
between the values measured at the beginning and on the third day
of culture containing only small cells (Fig.9a). By subtracting the
latter value (nuclear enzyme activity) from the enzyme activity of
cells stimulated for three days, by PHA the amount of the cytoplasmic DNA polymerase of PHA-stimulated cells can be obtained. Comparing the initial and maximal activities of the cytoplasmic DNA polymerase, a fourfold increase was obtained in the case of PHA-stimulated cells.

Experience with PHA has shown that both the incorporation of
^{3}H-thymidine and the activity of the DNA polymerase of tonsillar
lymphocytes can be increased under the conditions of culture applied. Therefore the decline of the DNA synthesis of non-stimulated
cells was due to some other, unknown circumstance and not to culture conditions.

ISOLATION AND CHARACTERIZATION OF DNA POLYMERASES

Freshly isolated tonsillar lymphocytes were shown to incorporate actively ^{3}H-thymidine, and the total DNA polymerase specific
activity of the cells was found to be higher than that of non-pro-
liferating blood lymphocytes. In order to prove that the high polymerase activity of the cells was due to the presence of the cytoplasmic enzyme that is characteristic of proliferating cells, we
separated the polymerase enzymes by chromatography.

Lymphocytes were homogenized in a hypotonic buffer solution.
Extraction with 1 M of KCl was performed for the solubilization of
the nuclear enzyme. The 100.000 g supernatant of the extract contained both the nuclear and the cytoplasmic enzymes. The supernatant was purified by DEAE cellulose chromatography. From the purified
preparation the cytoplasmic and the nuclear polymerases were separated by chromatography on a phosphocellulose column by gradient
elution with KCl (Fig.10). The α DNA polymerase was eluted with
0.25 M KCl; the β enzyme was eluted with 0.45 M KCl. The procedure
resulted in a 500-1.000 fold purification. Conditions of the assay
of enzyme activity are demonstrated in Table II. The reaction required the presence of divalent cations; Mn^{2+} could substitute Mg^{+2}
(Fig.11). The α polymerase was inhibited by NEM (N-ethylmaleimide)
and the β not.

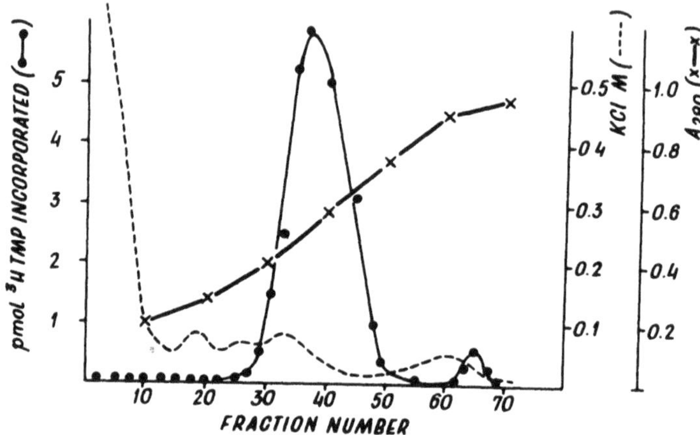

Fig. 10. Separation of DNA polymerases by phosphocellulose chromatography. Nucleic acids were removed from the 100.000 g supernatant of the homogenate of tonsillar cells by DEAE cellulose chromatography, afterwards the extract was applied onto a phosphocellulose column (1x0.9 cm). Gradient elution was performed by a buffer solution containing Tris-HCl, 0.05 M; pH 7.2; mercaptoethanolamine, 1 mM; glycerol, 20%; v/v, and with KCl as indicated (Antoni and Staub, 1978).

TABLE II

REQUIREMENTS FOR DNA SYNTHESIS

REACTION CONDITIONS	^3H-TMP INCORPORATED (pmol)	% CONTROL
COMPLETE	8.3	100
− DTT	7.1	86
− Mg^{2+}	0.5	6
− K^+	2.61	30
− TEMPLATE	0.26	3
− ENZYME	—	—
− d ATP	2.61	30
− d ATP, dCTP, dGTP	1.6	20

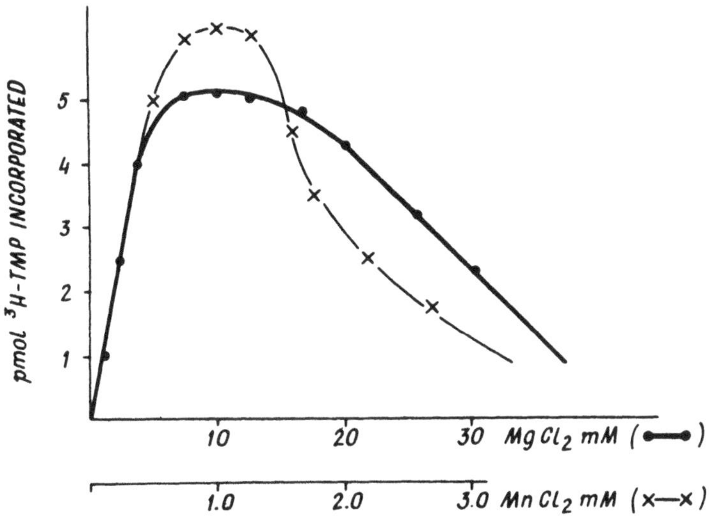

Fig. 11. Effect of Mg and Mn ions on the DNA polymerase reaction.
Assay mixture contained 50 mM Tris-HCl buffer, pH 7.8; dATP, dCTP,
dGTP, 0.12 mM each; 0.004 mM ³H-dTTP, (700 cpm/pmol); 0.4 mM KCL;
1.2 mM DTT; 50 µg calf thymus DNA, and enzyme solution containing
5-20 µg of protein and the salts as indicated. Final volume was
200 µl. The reaction was stopped by the addition of 0.3 ml 5% TCA.
Radioactivity of the precipitate was counted by a liquid scintillation spectrometer in toluene based cocktail.

In agreement with earlier data, the enzymological properties
of the two DNA polymerase enzymes were strikingly similar. The substrate and template saturation curves were identical for both enzymes, as well as the pH optima (between pH 7.4 and 8.0). The sim-
ilarity between the complex, high-molecular-weight cytoplasmic enzyme and the nuclear DNA polymerase, consisting of a single polypeptide chain, have been the subject of many speculations. The constant level and the individual immunological properties appear to
distinguish the nuclear enzyme from the cytoplasmic DNA polymerase.
However, experiments on sea urchin eggs have suggested that the
cytoplasmic enzyme is transferred into the nucleus in the early S
phase (Loeb et al. 1969). The translocation of an enzyme of several
hundred thousands of dalton molecular weight is hard to imagine,
but it is conceivable that subunits of the cytoplasmic DNA polymerase may gain access to the nucleus, the very site of DNA replication.
The latter assumption is in agreement with the similar catalytic
properties of the two polymerase enzymes, and it offers an explanation for the difference in their molecular weight. Further, the
variation in their sensitivity with respect to SH-blocking agents
can also be interpreted in these terms.

The interactions of the enzymes with the nuclear chromatin have not been elucidated either. It has been demonstrated that the nuclear DNA polymerase is firmly attached to the chromatin in the resting phase, but we do not know the properties of the chromatin-enzyme complex during the synthetic phase. One might presume that at that time a part of the loosely bound nuclear enzymes may leak into the cytoplasm. The chromatin-polymerase interaction has to change in the course of the cell cycle in another respect, too, i. e. the nuclear enzyme must be activated by some mechanism at the beginning of the S phase. It is supposed that this involves the modification (phosphorylation) of the enzyme. Most probably cAMP and cGMP are involved in the regulation of these processes.

REFERENCES

Anker, P., Stroun, M. and Maurice, P.A. Cancer Res. 36, 2832-2839 (1976).

Antoni, F. and Staub, M. Eds. Tonsils; Structure, Immunology and Biochemistry. Akadémiai Kiadc, Budapest (1978).

August, C.S., Merler, E., Lucas, D.O. and Janeway, C.A. Cell. Immunolog. 1. 603 (1970).

Baltimore, D. Nature (Lond.) 248, 409 (1974).

Bernheim, J.L., Mendelsohn, J., Kelley, M.F. and Dorian, R. Proc. Natl. Acad. Sci. USA 74, 2536-2540 (1977).

Bolden, A., Pedrali Noy, G. and Weissbach, A. J. Biol. Chem. 252, 3351 (1977).

Brown, G. and Greaves, M.F. Europ. J. Immunol. 4, 302 (1974).

Burgess, R.R. Procedure in Nucleic Acid Research, Vol.2. Cantoni, G.L. and Davics, D.R. (Eds.) Academic Press, New York.

Burton, K. Methods in Enzymology (L. Grossmann and K. Moldave, Eds.) Vol.12B, 163, Academic Press, New York (1968).

Chang, I.M.S. and Bollum, F.J. J. Biol. Chem. 246, 909 (1971).

Chang, I.M.S. and Bollum, F.J. J. Biol. Chem. 247, 7948 (1972).

Coleman, M.S., Hutton, J.J., De Simone, P. and Bollum, F.J. Proc. Nat. Acad. Sci. (Washington) 1, 4404 (1974).

Craig, R.K., Costello, P.A. and Keir, H.M. Biochem. J. 145, 233 (1975).

Demidenko, O.E., Tsvetkova, S.E. Bulletin of Exp. Biol. and Med. 83, 551-553 (1977).

Faragò, A., Antoni, F. and Fàbiàn, F. Biochem. Biophys. Acta (Amst.) 370, 459 (1974).

Fridlender, B., Fry, M., Bolden, A. and Weissbach, A. Proc. Nat. Acad. Sci. (Washington) 69, 452 (1972).

Gatien, J.G., Schnesberger, E.E., Parkman, R. and Merler, E. Europ. J. Immunol. 5, 306 (1975).

Geha, H.S. and Merler, E. Europ. J. Immunol. 4, 193 (1974).

Greaves, D.C. and Brown, G. J. Immunol. 112, 420 (1974).

Hoesli, D.C., Jones, A.P., Eisenstadt, J.M. and Wakslan, B.N. Int. Archs. Allergy Appl. Immun. 54, 517-526 (1977).

Holmes, A.M., Heslswood, J.D. and Johnston, J.R. Europ. J. Biochem. 43, 487 (1974).

Izui, S., Lambert, P.H., Fornié, G.J., Türler, J., Miescher, P.A. J. Exp. Med. 145, 1115-1130 (1977).

Lewis, B.J., Abrell, J.W., Smith, R.G. and Gallo, R.C. Science, 183, 867 (1974a).

Lewis, B.J., Abrell, J.W., Smith, R.G. and Gallo, R.C. Biochem. Biophys. Acta (Amst.) 349, 148 (1974b).

Loeb, L.A., Fransler, B., Williams, R. and Mazia, D. Exp. Cell. Res. 57, 298 (1969).

Lowry, O.H., Rosenbrough, N.J., Farr, A.L. and Randall, R.J. J. Biol. Chem. 193, 265 (1951).

Magnusson, G., Craing, R., Narkhammar, M., Reichard, P., Staub, M. and Warner, H. Cold Spring Harbor Symp. Quant. Biol. 39/2, 227 (1974).

Matsukage, A., Bohn, E.W. and Wilson, S.H. Proo. Nat. Acad. Sci. (Washington) 71, 578 (1974).

McCaffrey, R., Smoller, D.F. and Baltimore, D. Proc. Nat. Acad. Sci. (Washington) 70, 521 (1973).

Merler, E. and Silberschmidt, M. Immunology 22, 281 (1972).

Meuth, M., Aufreiter, E., Reichard, P. Eur. J. Biochem. 71, 39-43 (1976).

Moroni, C. and Schuman, G. Nature, 269, 601-602 (1977).

Piffkò, P., Köteles, G.J. and Antoni, F. Pract. Oto-rhing-laryng. 32, 350 (1970).

Rogers, J.G., Boldt, D., Kornfeld, S., Skinner, Sr.A. and Valeri, G.R. Proc. Natl. Acad. Sci. USA, 69, 1685-1689 (1972).

Rogers, J.C. Proc. Natl. Acad. Sci. USA, 73, 3211-3215 (1976).

Reichardt, P. Fed. Proc. 37, 9 (1978).

Sarin, P.S. and Gallo, R.C. J. Biol. Chem. 249, 8051 (1974).

Seifert, W.E. and Rudland, P.S. Nature (Lond.) 248, 138 (1974).

Smith, R.G. and Gallo, R.C. Proc. Nat. Acad. Sci. (Washington) 69, 2879 (1972).

Spadari, S. and Weissbach, A. J. Molec. Biol. 86, 11 (1974a).

Spadari, S. and Weissbach, A. J. Biol. Chem. 249, 5809 (1974b).

Stambrook, P.J. and Sisken, J.E. Biochem. Mol. 15, 246 (1976).

Staub, M. 9th FEBS Meeting, Budapest, Abstracts of Communications, p.163 (1974).

Staub, M., Warner, H.R. and Reichard, P. Biochem. Biophys. Res. Commun. 46, 1824 (1972).

Staub, M., Antoni, F. and Sellyei, M. Biochem. Med. 15, 246 (1976).

Staub, M., Faragò, A. and Antoni, F. Physical and Chemical Bases of Biological Information Transfer, First International Colloquium, Varna, Plenum Publ. Corp. New York, p.409 (1975).

Staub, M., Sasvàri-Székely, M., Spasokukotskaja, T., Antoni, F. and Merétey, K. Biochem. Med. 19, 218 (1978a).

Staub, M. and Antoni, F. Nucleic Acids Res. $\underline{5}$, 3071 (1978b).
Tabata, T., Enomoto, T., Fujimura, N. and Riramatsu, K. Acta oto-
    laryng. (Stockholm) $\underline{77}$, 150 (1974).
Tyrsted, G. and Munch-Petersen, B. Nucleic Acids Research, $\underline{4}$,
    2713 (1977).
Weissbach, A. Ann. Rev. Biochem. $\underline{46}$, 25 (1977).

THE EFFECT OF UV LIGHT ON THE HUMAN LYMPHOCYTES

F. Antoni, I. Csuka, I. Vincze, Gy. Farkas, and M. Staub

1st Institute of Biochemistry, Semmelweis University

Medical School, Budapest, Hungary

INTRODUCTION

Lymphocytes can be separated by different methods into two
subpopulations: T cells (the thymus derived cells) and B cells
(bone marrow type lymphocytes) (1,2,3).

These two cell types develop from the same stem cell and they
differentiate during the maturation. The receptors on the cell surface of T and B cells are different and they have different roles
in the immune reaction.

The sensitivity of T and B cells against UV light is different
as has been reported by Horowitz and coworkers (4). Painter (5)
supposed that the T lymphocytes have a lower repair capacity than
B cells.

The UV light produces different damages in the cell. The DNA
damages and the repair of these injuries are investigated in detail
but we have very little information about the other UV light provoked disturbances in the cell.

At the lymphocytes taking part in the humoral and cell mediated
immune reactions the UV light induced cell membrane damage seem to
present an important problem.

These damages may occur in proteins having structural and
transport functions or in proteins taking part in the immune processes such as antigens, immunoglobulins, receptors.

Starting from these considerations we have studied the DNA

repair synthesis in isolated B and T lymphocytes and the effect of
UV light on the red blood cell binding receptors of T cell as well
as the c'3 receptors of B cells.

MATERIALS AND METHODS

All of the used inorganic materials were the products of
REANAL.

Nucleotides triphosphates and calf thymus DNA were purchased
from Calbiochem.

Methyl-^{3}H-thymidine (20 Ci/mmol) was the product of Institute
for Research Production and Utilisation of Radioisotopes, Prague,
C.S.S.R.

Lymphocytes were prepared from human tonsils by the method of
Antoni (6). The tonsils were cut into small pieces and shaken in
Hank's medium containing 100 µg streptomycin and 100 I.U. penicyl-
lin. The suspension was filtered through four layers of gauze and
the cells were sedimented by centrifugation. The cells were washed
with Earle's medium and finally they were suspended in Earle's medium
containing 2% fetal calf serum.

The T and B cells were separated by using wool column according to Greaves and Brown (7). The cross-contamination between the
separated T and B lymphocyte fractions was below 15% as judged on
the basis of E and EAC rosette forming ability.

The cell suspensions were irradiated using a Tungsram germicide
lamp (300 erg/mm^2/min).

The nucleotides were extracted by the method of Vincze et al.
(8). The cells were incubated for 30 minutes in Hank's medium containing 10 µci/ml ^{3}H-thymidine and $5 \cdot 10^{-3}$ M hydroxyurea to label the
pool and to reproduce the condition of repair synthesis. After the
incubation the cells were centrifuged and the pellet was suspended
in distilled water. The suspension was frozen and melted three times
to disrupt the cells and it was mixed with an equal volume of ethanol to precipitate the protein, DNA, RNA. After centrifugation,
the clear supernatant was pooled and dried by lyophylisation and
from the pellet the DNA was determined. DNA was determined according to Burton (9).

The in vitro DNA polymerase reaction system contained 25 mM
Tris-HCl buffer pH 7.4, 2 mM KCl, 4.4 mM MgCl$_2$, 25 µg activated oal
thymus DNA, DNA-polymerase enzyme 100 µg in a volume of 0.2 ml.

DNA was activated according to Aposhian and Kornberg (10).
DNA polymerase was isolated from human lymphocytes by the methods
of M. Staub (11).

Radioactivity was measured by liquid scintillation counting
and the activity was expressed in dpm/µg DNA.

The red blood cell binding receptors were determined by the
E rosette forming ability of T lymphocytes following the method of
Jondall et al. (12).

The c'3 receptors of B cells were determined by EAC rosette
forming ability of the cells according to Bianco et al. (13).

The viability of the cells was followed by the tripane blue
exclusion test according to Phillips (14).

RESULTS

T and B cells suspended in Hank's medium were preincubated at
0°C for 10 minutes with hydroxyurea and irradiated with 100 erg/mm^2.
After irradiation the cell suspensions were completed with ^{3}H-thym-
idine and incubated at 37°C. Control cells, not irradiated with UV
light, were incubated under similar conditions. Dpm in ^{3}H-thymidine
incorporated by irradiated cells was corrected for the dpm value of
control cells and plotted against the incubation time. In this way
straight lines were obtained representing the incorporation induced
by UV light in B and C cells (see Fig.1).

As it is shown in Fig.1, during the repair synthesis B cells
incorporated twice as many thymidine than T cells. The capacity of
the DNA repair synthesis was determined using UV light irradiation
from 25 to 400 erg/mm^2. After 30 minutes incubation the UV light
induced incorporation of labeled thymidine was measured. Plotting
the incorporation values as a function of UV light dose two curves
were obtained (see Fig.2). These curves show a similar pattern however the specific radioactivities measured with T cells were found
to be about twofold smaller as compared to the respective values
of the B cells.

The rate of the incorporation of ^{3}H-thymidine into DNA depends
on several factors such as the number of initiation sites for the
DNA polymerase and the activity or level of the polymerase enzyme,
the size of the nucleotide pool and especially the specific radioactivity of thymidine triphosphate in the nucleotide pool. The
ability of this pool to support the incorporation of labeled thymidine could be estimated by the addition of the nucleotides extracted from the cells to an in vitro DNA polymerase reaction system.

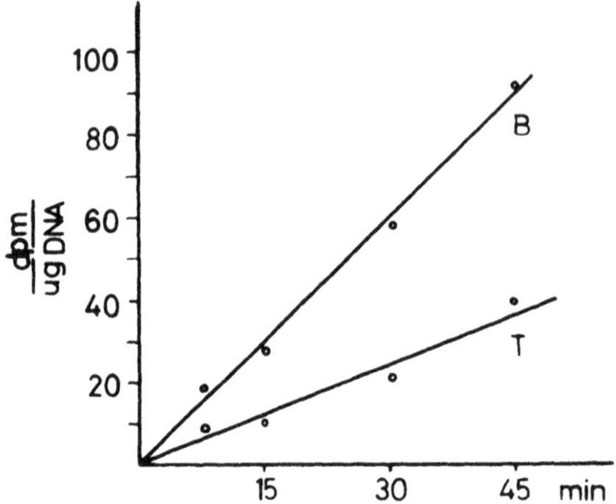

FIGURE 1. The UV light induced incorporation of T and B lymphocytes as a function of incubation time. Suspensions of T and B
cells (10^7 cells/ml) were preincubated for 10 minutes at $0°C$ in
Hank's medium containing $5 \cdot 10^{-3}$ M hydroxyurea, 100 I.U. penicillin
and 100 µg streptomycin. Cells were irradiated with 100 erg/mm^2,
^{3}H-thymidine was added (5 µCi/ml) and the suspensions were incubated
for the times indicated on the abscisse. Incubation was stopped by
adding 3 ml of ice cold 0.5 N PCA to the 1.0 ml cell suspensions.
The pellets were washed four times with 0.5 M PCA and after hydro-
lysis DNA and radioactivity were determined from separated aliquots.

In order to carry out this experiment isolated T and B cells
were extracted as described in Materials and Methods. The dry extract was dissolved in distilled water and different aliquots were
added to separate tubes containing the polymerase reaction system.

As it is shown in Fig.3 in this way we obtained straight lines.
Since the amount of the extracts were calculated on a DNA basis a
comparison of the slopes of the lines permits the conclusion that
the pool of the B cells allowed a two times higher incorporation
compared with the value of T cell pool. This difference between B
and C cells could be either due to a twofold difference in the
specific radioactivity of thymidine triphosphate or due to a two-
fold difference in the concentration of some rate limiting nucleo-
tide(s).

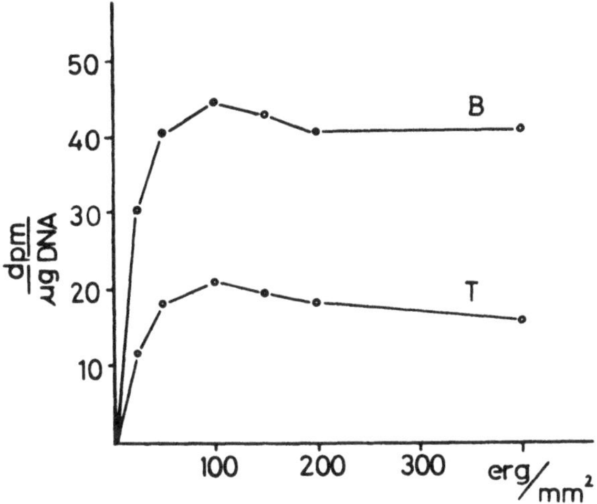

FIGURE 2. DNA repair synthesis as a function of UV light dose. The condition of the experiment same as at Fig.1. The cells were irradiated with UV light dose indicated on the abscisse and the incubation time was 30 minutes after the addition of ^{3}H-thymidine.

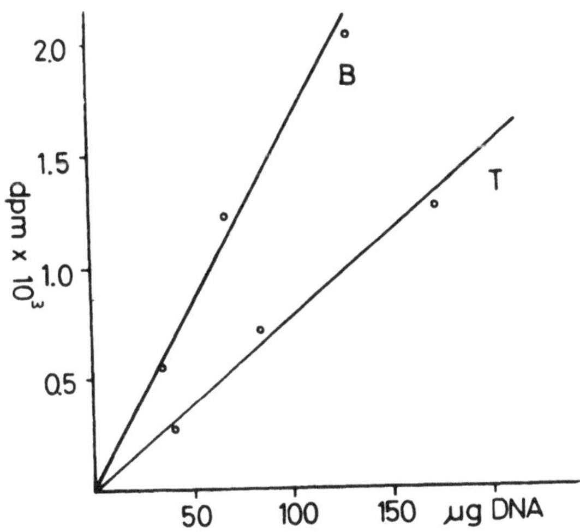

FIGURE 3. Incorporation of ^{3}H-thymidine from the nucleotides of T and B cells. The condition of the reaction seen in Material and Methods.

The effect of UV light in the membrane receptors of T and B
cells was studied immediately after the irradiation with different
UV light dose between 50 and 400 erg/mm^2.

As it is shown in the Fig.4, the UV light irradiation markedly
decreases the erythrocyte binding receptors of T cell, that is, the
E rosette formation, but it has a very small effect on the c'3 receptors of B cells as can be judged from the EAC rosette forming
ability of this cell type.

We have determined parallel the viability and rosette forming
ability of T and B cells to solve the question whether or not the
change of the rosette formation is a consequence of the change in
the viability of the cells.

As shown in Fig.5 contrary to the rosette forming ability the
viability of T cells decreases only to a very small extent, when
determined immediately after the irradiation.

This dose dependent decrease of the rosette formation can be
divided into two parts. Up to 100 erg/mm^2 the decrease is propor-
tionate to the dose of UV light and is of a very great extent. Increasing the UV light dose the rosette formation decreases further
to a smaller extent.

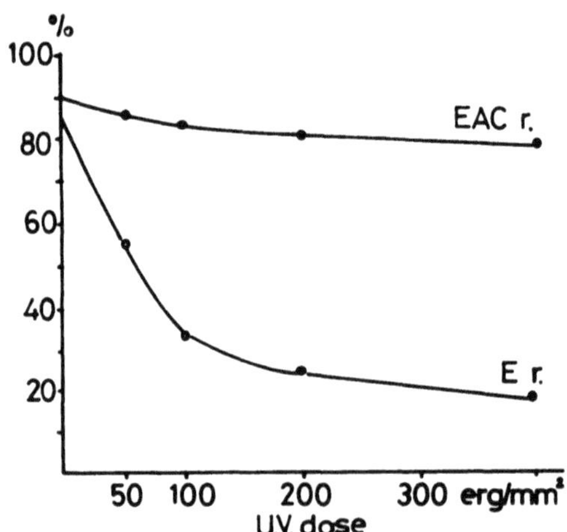

FIGURE 4. The effect of UV light on the E and EAC formation
at separated human T and B lymphocytes. The lymphocytes were irradiated in Hank's medium with different UV light dose as indicated
on the abscisse. The determination of E and EAC rosette formation
see in Material and Methods. Er means the percent of E rosette forming cells among the isolated T lymphocytes. EAC r. means the percent
of EAC rosette forming cells among the isolated B lymphocytes.

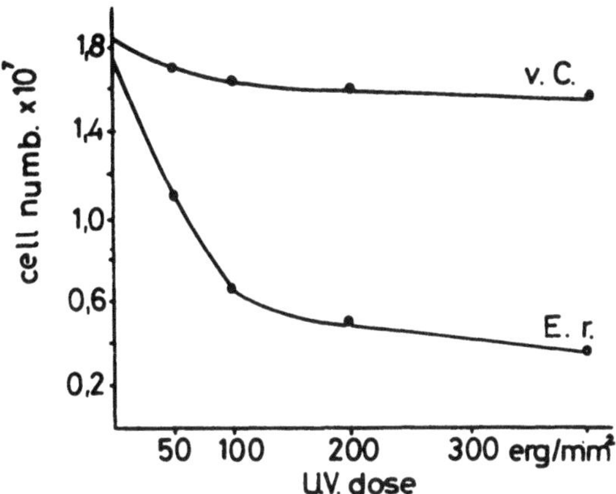

FIGURE 5.   The effect of UV light on the viability and E rosette forming ability of isolated T cells. The condition of the experiment same as at Fig.4. The determination of viable cells see in Material and Methods. v.C. means the number of viable cells. E.r. means the number of cells forming E rosette.

From these data it can be concluded that from the point of view of the receptor sensitivity against UV light two subpopulations of T cells occur.

The c'3 receptors of B cells are more resistant to UV light irradiation, however, from the data of Fig.6 a similar conclusion can be reached that is two subpopulations of B cells occur according to the sensitivity of c'3 receptors.

The decrease of the E rosette forming ability of T cells that is the inactivation of the red blood cell binding receptors is not only a rapid process occuring during or immediately after the irradiation a further decrease was found when the irradiated cells were incubated in Parker medium.

The rosette forming ability and viability of T cells was determined immediately and 1, 2, 12 hours after the irradiation with 50 and 200 erg/mm². During the incubation following the irradiation the viability of T cells decreased but the rosette forming ability of the cells decreased more rapidly as shown in Fig.7.

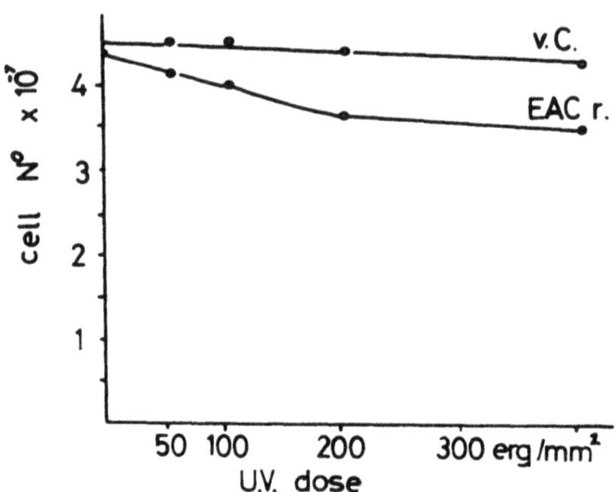

FIGURE 6.   The effect of UV light on the viability and EAC
rosette forming ability of isolated B cells. The condition of the
experiment same as in Fig.4. The determination of EAC rosette forming cells and the viability of the cells see in Material and Methods.
v.C. means the number of viable cells. EAC.r. means the number of
EAC rosette forming cells.

The rate of this time dependent inactivation of the red blood
cell binding receptors of T cells seems to be independent of the
dose of UV light. A similar rate of inactivation was found after
50 or 200 erg/mm² UV light dose.

As it is shown in Fig.8, the decrease of the viability and
EAC rosette forming ability of B cells are the same during the irradiation followed incubation. This means that the time dependent inactivation of the c'3 receptors of B cells did not occur. Comparing
the viability of T and B cells during the irradiation followed incubation we have not found significant difference between the two
lymphocyte populations as shown by the data of Fig.9.

DISCUSSION

According to the data presented here it can be concluded that
the sensitivity of T and B cells to UV light are the same (Fig.9).
The ability of the T and B cells to repair the UV light produced
DNA damage is apparently different (Fig.1 and 2) but this is a consequence of the differently labelled nucleotide pools (Fig.3).

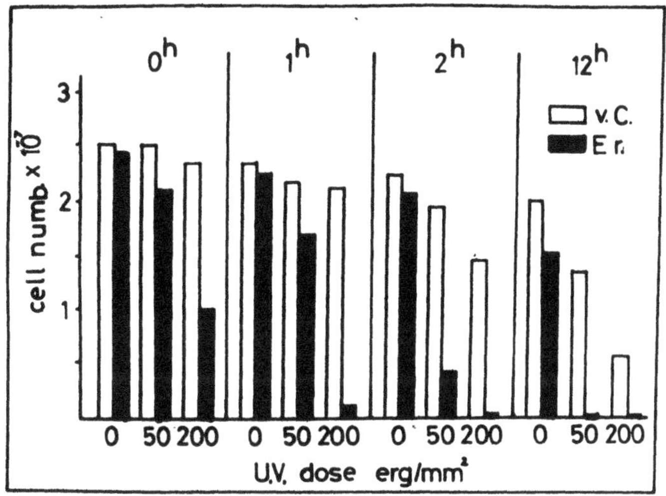

FIGURE 7. The change of viability and E rosette forming
ability of T cells during the irradiation followed incubation. The
isolated T cells were irradiated in Hank's medium with 50 and 200
erg/mm and incubated in Parker medium. Aliquotes were taken to measure the viability (v.C.) and rosette forming ability (E.r.) of the
irradiated cell after the irradiation immediately (0h) 1,2, 12 hours.
0 means the date of the control cells. 50 and 200 means the irradiated cells.

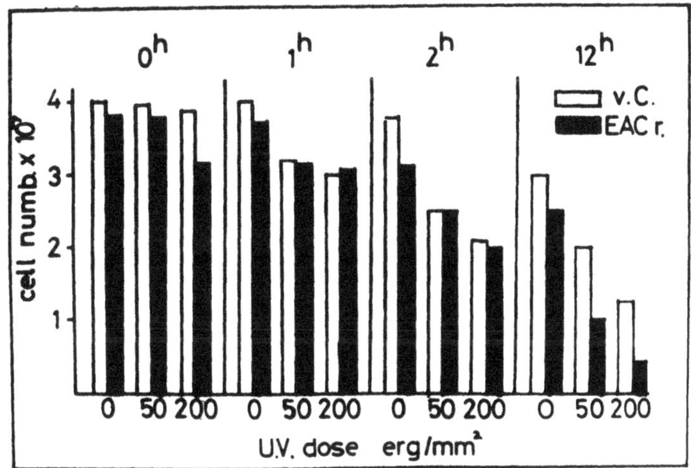

FIGURE 8. The change of the viability and EAC rosette forming
ability of B cells during the irradiation followed incubation. The
condition of the experiment same as in Fig.7.

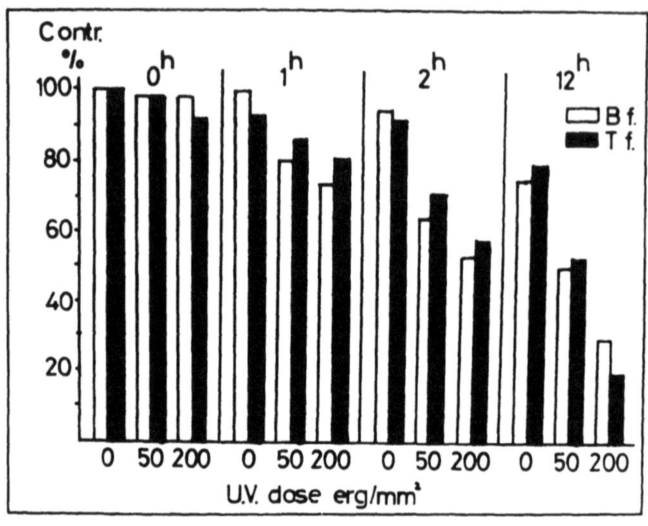

FIGURE 9. The change of the viability of T and B cells during
the irradiation followed incubation. The condition of the experiment
same as in Fig.7. B.f. means the viable cells of the B cell fraction.
T.f. means the viable cells of the T cell fraction.

The red blood cell binding receptors of T cells are very UV
light sensitive compared to the c'3 receptors of B cells (Fig. 4).
Horowitz and co-workers (4) have determined the viability of T and
B cells comparing the E rosette and EAC rosette forming ability of
the viable cells in a mixed T and B lymphocyte population. Since
the E rosette formation was very UV light sensitive compared to
the EAC rosette formation they suggested that the T cells are more
sensitive to UV light. Using isolated T and B lymphocytes we could
not support this suggestion (Fig.5, 6, 9).

The inactivation of the red blood cell binding receptors of T
cells and the c'3 receptors of B cells during or immediately after
the irradiation may be a photooxydation or other photochemical reaction of the receptor proteins. Another possibility may be that
the photochemical reaction of the lipid bilayer around the receptors
induces a change in the position of the receptor in the cell membrane lipid bilayer by this way inhibiting it in its function. The
time dependent and apparently UV light dose independent inactivation
of red blood cell binding receptors of T cells (Fig.7) are possibly
a consequence of the UV light provoked reorganisation of the cell
membrane. During this process the cells lose these receptors or the
receptors accumulate at one part of the cell membrane, so called

cap formation (15), and so they cannot bind but one or two red blood cells, however, the rosette formation does not occur.(16,17).

The UV light induced reorganisation of the B cell membrane could not be detected (Fig.8), or else the c'3 receptors did not participate in this process.

REFERENCES

1. Geha, R.S., Rosen, F.S., and Merler, E., Nature 248, 426 (1974).
2. Bobrove, A.M., Strober, S., Herzenberg, L.A., and De Pamphilis, J.D., The Journal of Immunology, 112, 520 (1974).
3. Zeylemaker, W.P., Roos, M.Th.L., Meyer, C.J.L.M., Schellekens, P.Th.A., and Eijsvoogel, V.P., Cellular Immunology, 14, 346 (1974).
4. Horowitz, S., Cripps, D., and Hong, R., Cellular Immunology, 14, 80 (1974).
5. Painter, R.B. in "Photophysiology", Ed. Giese, A.C. Academic Press, New York, Vol.5, p.169 (1970).
6. Antoni, F. in "Tonsils, Structure, Immunology and Biochemistry", Eds. Antoni, F., Staub, M. Akadémiai Kiadò, Budapest, p.80 (1978).
7. Greaves, F.M., and Brown, G., The Journal of Immunology, 112, 420 (1974).
8. Vincze, I., Csuka, I., Farkas, Gy., Staub, M., and Antoni, F., in "DNA repair and late effects" Ed. Roetzer, Eisenstadt, p.73 (1976).
9. Burton, K., Methods in Enzymology (Edited by Grossman, L., Molcave, K.) XII. part B p.163 (1968).
10. Aposhian, H.V., and Kornberg, A., Journal of Biological Chemistry, 237, 519 (1962).
11. Staub, M., Information Transfer (Supl) Pergamon Press, International Colloquium, Varna (1975).
12. Jondal, M., Holm, G., and Wigzell, H.J., Exp. Med. 136, 207 (1972).
13. Bianco, C., Patrick, R., and Nussenzweig, V.J., Exp. Med. 132, 702 (1970).
14. Phillips, H.J., Dye exclusion test for cell viability, Academic Press, New York (1973).
15. Taylor, R.B., Duffus, W.P.H., Raff, M.C., and Petris, S., Nature New Biology, 233, 225 (1971).
16. Csuka, I., Farkas, Gy., Vincze, I., and Antoni, F., 12th Annual Meeting of European Society of Radiation Biology, Budapest, abs. 26 (1976).
17. Csuka, I., Vincze, I., Farkas, Gy., and Antoni, F., Acta Physiol. Acad. Sci. Hung. 52/2-3 (1978).

RADIATION SENSITIVITY OF STIMULATED HUMAN LYMPHOCYTES: RELATION

TO DNA REPAIR AND CELL MEMBRANE ACTIVATION PROCESSES

A. Castellani, C. Catena, G. Biondi°

Laboratorio di Radiopatologia, CNEN, CSN, Casaccia,Rome

°Cattedra di Radiobiologia, University of Siena, Italy

Human lymphocytes from peripheral blood are considered to be among the most radiosensitive cells in vivo (1) as well as in vitro (2-3).

In vivo the absolute number of lymphocytes present in the peripheral blood after irradiation is often taken as a sensitive biological indicator of radiation exposure (4-5).

Morphological changes in X-irradiated lymphocytes have been observed after doses of 0.25-2 Gy. The first changes are reflected in the appearance of vacuoles within the cell nucleus, disruption of the chromatin network ending with pyknosis and nuclear fragmentation (6). Destruction of the cells has been noted within 4-6 hours after irradiation.

The lymphocytes, non dividing cells in the G_o phase of the cell cycle, after stimulation with mitogens are triggered to enter the G_1 phase and become large lymphoblasts, much more radioresistant than unstimulated lymphocytes. An induced radioresistance in stimulated lymphocytes has been observed by Schrek and Stefani (7) who demonstrated that PHA-stimulated lymphocytes show a relatively high degree of radioresistance compared to the unstimulated ones.

Fig.1 shows the survival of human lymphocytes for increasing doses of X-ray in PHA-stimulated as well as in unstimulated ones. The time course of this phenomenon is represented in fig.2 (8) where it is shown that 500 rad given to unstimulated lymphocytes results in a decrease in survival reaching 10% survival after 40 hours from irradiation. This phenomenon is absent or strongly

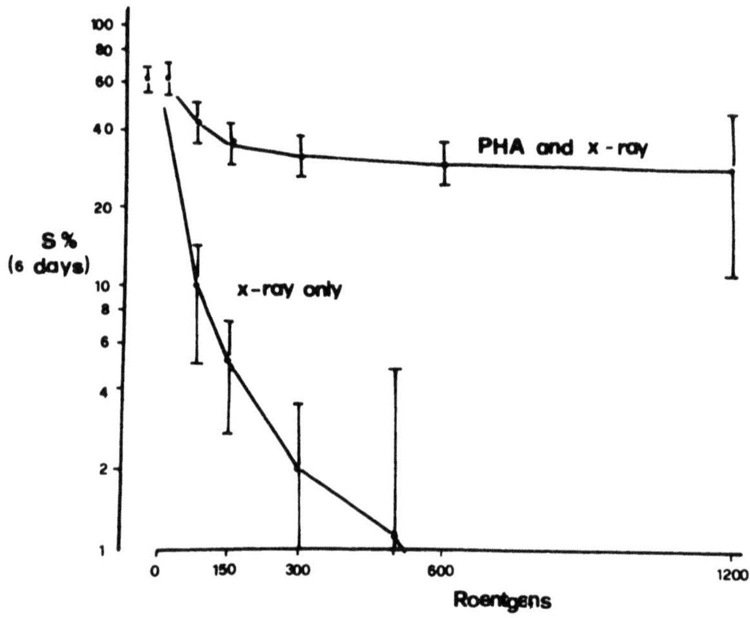

FIGURE 1.   Effect of PHA on survival of X-irradiated human
lymphocytes. Abscissa: doses of X-rays. Ordinate: percentage of
lymphocytes surviving after 6 days of incubation. (Schrek and
Stefani, 1958 - modified).

reduced in PHA-stimulated lymphocytes as shown in the upper curves
of the fig.2.

40 hours after PHA stimulation the lymphocytes begin to enter
the first mitosis (9-10). At that time the effect of irradiation
on PHA-stimulated lymphocytes appears as mitotic death. Sato reported that: "the addition of PHA protected small lymphocytes from
early normal interphase death, but changed the type of cell-death
into the one related to mitosis" (8). Clearly, PHA-induced biochemical changes in the cells are responsible for the changes in
radiation response.

In order to obtain information on the acquired radioresistance
of PHA-stimulated lymphocytes we have studied the DNA repair capability and a cell membrane function in irradiated PHA-stimulated
lymphocytes. Since in all our experiments the cells were irradiated in S phase, in order to have a clear knowledge of the experimental system to work with we first studied the growth kinetics of
PHA-stimulated lymphocytes to evaluate the kinetics of cell entry
into the first S phase. In fig.3 is reported the incorporation of

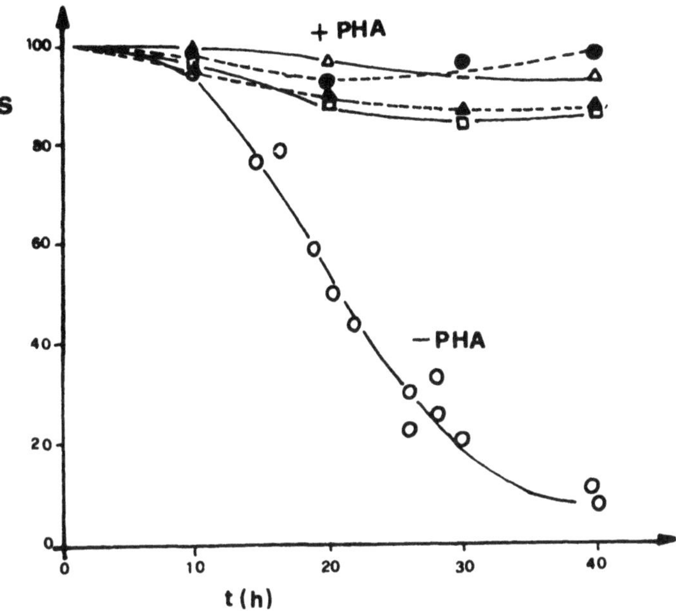

FIGURE 2. Time-course survival of X-irradiated human lymphocytes. Abscissa: culture time after irradiation. Ordinate: percentage of survival. (Sato, 1970 - modified).

[3]H-dThd in TCA insoluble cell fraction for 30 minutes pulses at
various time after PHA stimulation (11). Each point is related to
the increment in number of cells which, at given time, enter in S
phase. The curve in fig.3 is similar to that obtained by MacKinney
(12) who studied the increase of ^{3}H-dThd labelled cell in relation
to the time after the PHA stimulation. The time course of this
phenomenon is exponential up to 40 hours when the curve declines
possibly because of the exit of cells from S phase (11). To have
information on cell age within the S phase we assume that the cells
go through three phases before leaving S phase: early, medium and
late S phase. We also assume that each one of these phases is four
hours in length.

Fig.4 was obtained by displacement of the experimental curve,
along the abscissa, by 4, 8, or 12 hours. This graph gives an estimate of the percentage of cells of various ages within the S
phase at any given time. Our experiments were performed with lymphocytes irradiated in S phase from 36 to 48 hours, when the majority of the cell population is in the medium and late S phase.

We measured the ^{3}H-dThd uptake into soluble and insoluble
fractions of both irradiated and control cells, under experimental

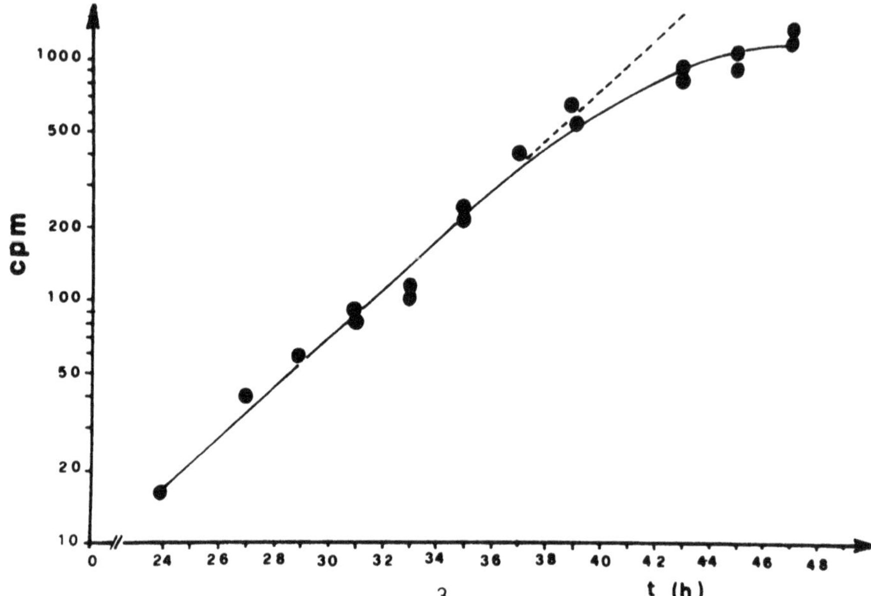

FIGURE 3. Time-course of ^{3}H-dThd incorporation in TCA insoluble fraction in human lymphocytes at various time from PHA stimulation.

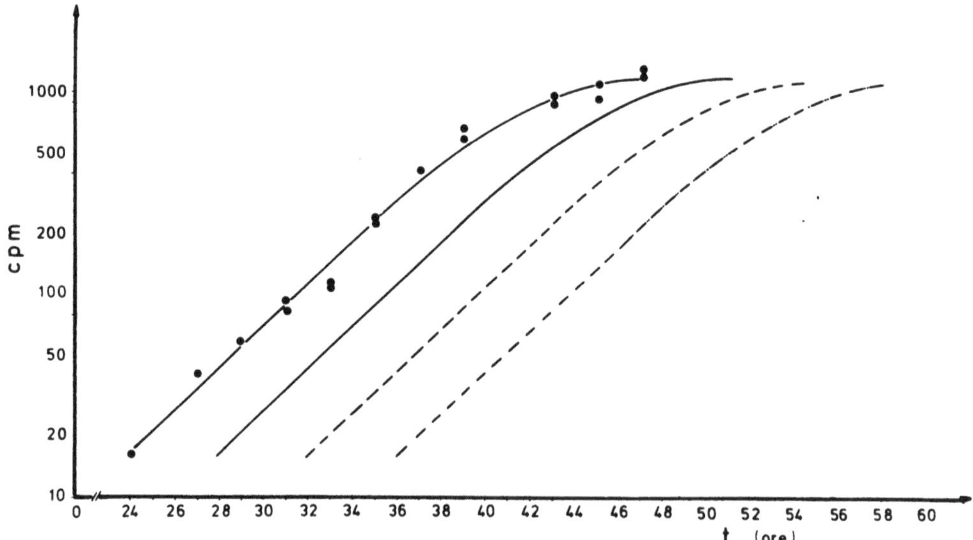

FIGURE 4. Kinetic of entry in S phase of PHA-stimulated lymphocytes. (-) early S phase, (-.-) medium S phase, (-..-) late S phase.

conditions which allow us to evaluate the extent of the damage to membrane function. Fig.5 shows the time-course of ^{3}H-rUrd and ^{3}H-dThd uptake in PHA-stimulated lymphocytes.

A drastic increase of ^{3}H-dThd uptake via mediated transport (13-15) is observed in S phase cell, starting 24 hours after PHA stimulation (at G_1-S phase transition), while the ^{3}H-rUrd uptake increases in the early hours after stimulation (G_1-phase). The pool is the difference between the total uptake and the acid insoluble fraction.

Fig.6 shows the saturation curves for the uptake of thymidine by PHA-stimulated human lymphocytes measured 46 hours after PHA

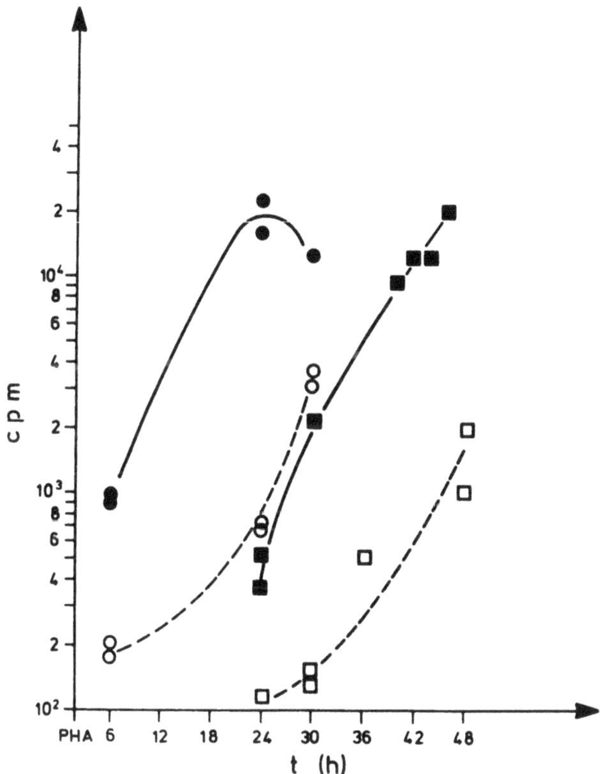

FIGURE 5. Time-course of the uptake of ^{3}H-rUrd and ^{3}H-dThd in human lymphocytes at various times from PHA stimulation.
● ^{3}H-rUrd sol. + insol.
O ^{3}H-rUrd TCA insol.
■ ^{3}H-dThd sol. + insol.
□ ^{3}H-dThd TCA insol.

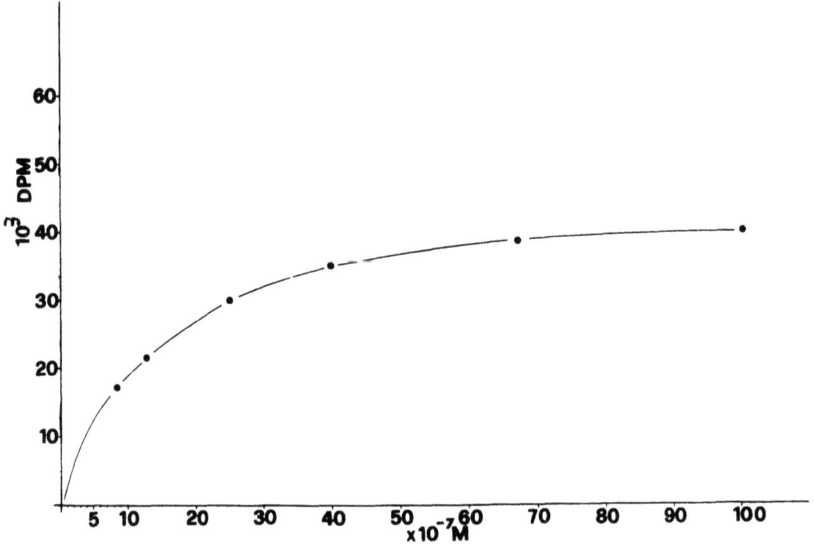

FIGURE 6. Saturation curve for the uptake of ^{3}H-dThd by
human PHA-stimulated lymphocytes. Abscissa: ^{3}H-dThd concentration
in external medium. Ordinate: radioactivity (dpm) in the whole
cell.

stimulation. These results strongly support the existence of a
saturable carrier-transport system of dThd in human lymphocytes,
similar to that shown in Ehrlich ascites tumour cells (16).

We were mainly interested in the variation of these uptake
properties of lymphocytes, caused by irradiation.

Fig.7 shows dose-effect curve for ^{3}H-dThd uptake in PHA-
stimulated lymphocytes, for thymidine concentration in the medium
corresponding to the saturation values.

This curve shows the radioactivity in the cellular pool which
was obtained subtracting the radioactivity in the TCA insoluble
fraction from the radioactivity in the total cell material.

The curve is a multiphase curve. A shoulder is evident up to
10 Gy. The membrane radioresistance induced by PHA treatment allows the cells to avoid interphase death and to proceed in the
cell cycle entering DNA synthesis. We can measure the DNA synthesis in irradiated cells and eventually look for repair of the
radiation damage.

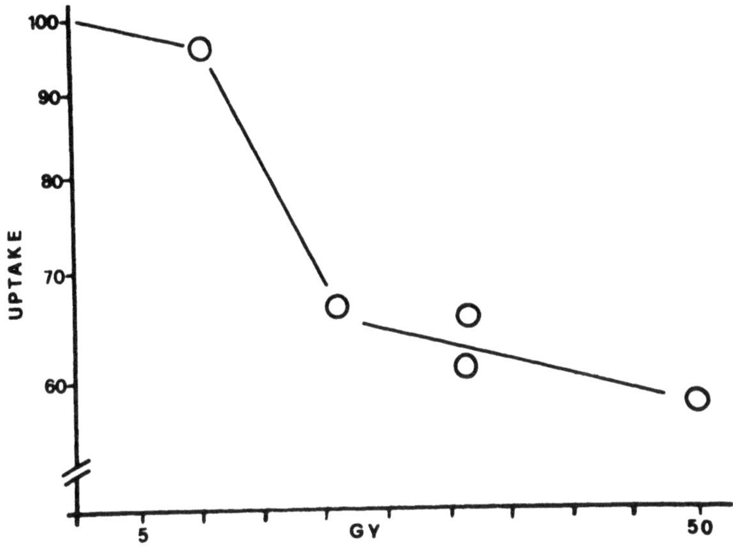

FIGURE 7. Dose-effect curve for ^{3}H-dThd uptake in human X-irradiated lymphocytes.

Fig.8 gives some indication of the repair process which occurs in lymphocytes irradiated 40 hours after the PHA stimulation. The upper curve shows the DNA synthesis in control cells at various times after PHA stimulation for 30 min pulse with ^{3}H-dThd. The lower curve was obtained measuring the DNA synthesis immediately after irradiation and at various time intervals following the irradiation. The effect of 90 Gy results in 70% DNA synthesis depression if measured immediatly after irradiation. This depression decreases with time. After 6 hours, the values of thymidine incorporation approach the values of the control.

Further indications on DNA repair synthesis in irradiated lymphocytes can be obtained with the use of hydroxyurea which, in a given experimental condition (2 10^{-3}M for 30 minutes), inhibits the semiconservative DNA synthesis leaving repair synthesis unaffected (17-18-19-20-21). Fig.9 (A) shows the inhibition of DNA synthesis in PHA-stimulated human lymphocytes irradiated with 25 Gy during the S phase (22).

In the same figure is also shown the inhibitory effect of HU 2.10^{-3}M in the control cell (B). The last column (C) shows the inhibitory effect of HU in irradiated cells that is the combined

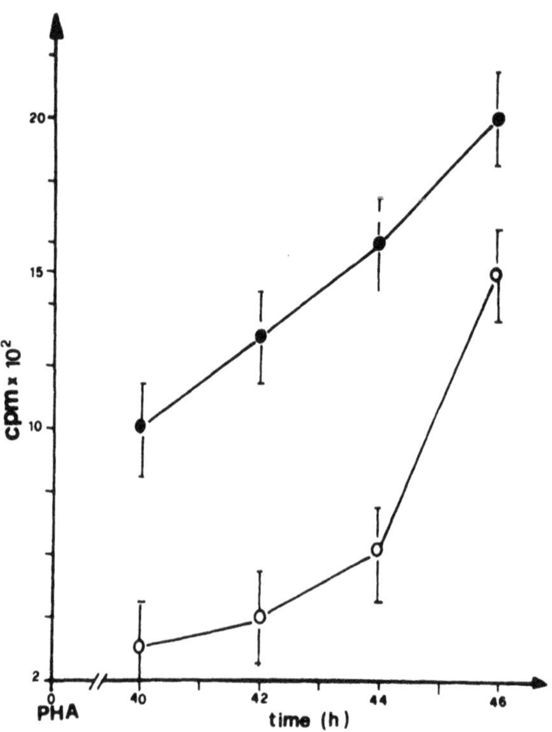

FIGURE 8. ^{3}H-dThd incorporation in TCA insol. fraction in
control (•) and in X-irradiated at 40 h with 90 Gy (o) human
lymphocytes. Abscissa: time (h) from PHA stimulation. Ordinate:
radioactivity in TCA insoluble fraction.

effect of HU and radiation.

Radiation alone gives about 20% inhibition, HU alone gives
40% inhibition while the combined treatment gives 25% inhibition.
The effects were measured immediately after irradiation. This
suggests that immediately after irradiation the residual DNA
synthesis is partially insensitive to HU.

Fig.10 shows the inhibitory effect of HU on residual DNA synthesis in irradiated cells for HU treatment carried out at various
times after irradiation. The white columns represent theoretical

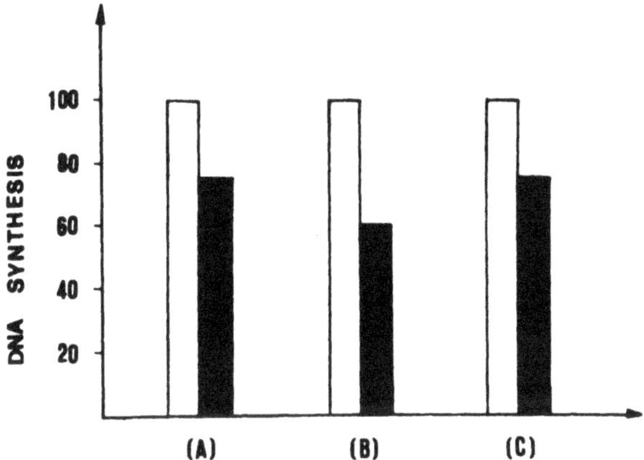

FIGURE 9. DNA synthesis inhibition in PHA-stimulated human lymphocytes treated with: A) 25 Gy B) HU ($2 \cdot 10^{-3}$ M for 30 min) C) 25 Gy + HU ($2 \cdot 10^{-3}$ M for 30 min). The open bars are untreated samples, the solid bars are treated samples.

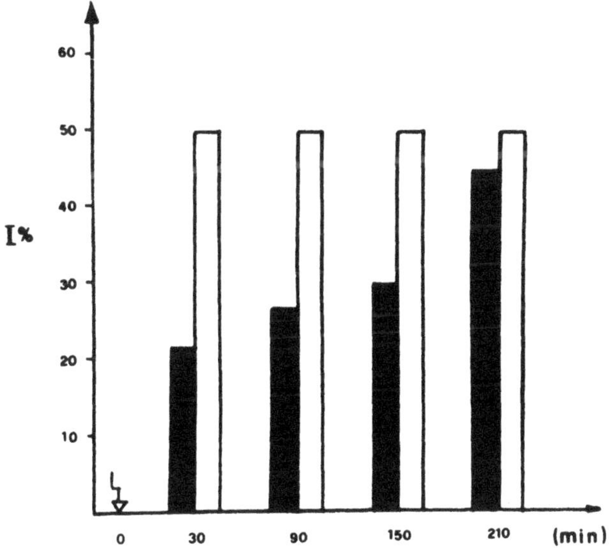

FIGURE 10. Effect of HU on residual DNA synthesis in human lymphocytes X-irradiated (25 Gy) at 40 hours from PHA stimulation. White column: theoretical values; black column: experimental values. Abscissa: time (minutes) after X-irradiation. Ordinate: percentage inhibition of DNA synthesis.

values obtained by simple summation of DNA synthesis inhibition by
irradiation plus HU inhibition expressed as 40% of the residual
DNA synthesis assuming to be of semiconservative type. The black
columns represent the experimental values. It is evident that the
sensitivity of the residual DNA synthesis becomes progressively
sensitive to HU, indicating that by 210 min the repair is almost
completed.

DISCUSSION

The radiation biology of human lymphocytes is dominated by
two phenomena, namely: a) high radiosensitivity of lymphocytes
which die in interphase, b) the PHA-induced relative radioresistance of these cells which, after stimulation, escape the interphase death and eventually die in mitosis. We think that these
phenomena constitute a good system to work with in order to study
some of the factors which control the response of human cells to
radiation.

The purpose of our study was to correlate the development of
the relative radioresistance in PHA-stimulated lymphocytes with
the transformation process, particularly with membrane activation
and increasing repair activity in human lymphocytes in S phase.
There is a certain amount of experimental evidence which suggests
that interphase death is correlated with modifications of structural and functional integrity of the cell membrane. Recently Sato
and coll. (23-24) have demonstrated that PHA treatment prevents
loss of lymphocytes membrane negative charges by radiation up to
a dose of 5 Gy. PHA works in blocking the very early conformational changes in cell membrane induced by radiation.

We have studied the radiosensitivity of the thymidine transport system of the lymphocytes cell membrane, perhaps not directly
connected with cell survival but representative of a very specific
funtion of this cell membrane. Our data demonstrate the existence
of a carrier-mediated transport for thymidine across the human
lymphocytes cell membrane. This transport appears not to work in
unstimulated lymphocytes and is activated in G_1-S phase transition
as Peters and Hausen (13) found in bovine lymphocytes. We have
studied the radioresistance of the thymidine transport not only as
a specific function, operating in S phase, but also because any
eyentual impairment of this function could affect the measure
(^{3}H-dThd uptake) of DNA synthesis in irradiated cells.

The dose-response curve for ^{3}H-dThd transport in S phase PHA-
stimulated lymphocytes shows a shoulder up to 10 Gy which is representative of a radioresistance of this cell membrane function.

These data also account for the fact that, within this dose range, the ^{3}H-dThd transport does not influence the depression of DNA synthesis by X-ray. The irradiated cells progress into the cell cycle, stimulated lymphocytes enter the S phase and eventually undergo mitosis.

The experiments presented in this paper demonstrate that DNA repair phenomena are active in S phase PHA-stimulated lymphocytes. This is, probably, the second factor which may be responsible for the radioresistance of stimulated lymphocytes.

DNA repair capability allows the cells correctly to complete DNA replication and decreases the probability of mitotic death.

Recently it has been proposed (25-26-27) the use of an HU test and particularly the analysis of the repair saturation curves in order to gain information on the repair capability of radiation damage in human lymphocytes from various individuals who are professionally exposed.

REFERENCES

1. Globa, S., Globa, M., Wilczok, T., Int. J. Radiat. Biol. 31, 261-268 (1967).
2. Schrek, R., J. Lab. Clin. Med. 51, 904-915 (1958).
3. Durum, S.H., Gengozian, N., Int. J. Radiat. Biol. 34, n.1, 1-15 (1978).
4. Vos, O., Effects of Ionising Radiation on the Haematopoietic Tissue, I.A.E.A., Vienna, 134 (1967).
5. Dienstbier, Z., Arient, M., Pospil, J., Effects of Ionising Radiation on the Haematopoietic Tissue, I.A.E.A., Vienna, 134 (1967).
6. Scaife, J.F., Brohée, H., Int. J. Radiat. Biol. 13, 111-136 (1967).
7. Schrek, R., Stefani, S., J. Nat. Cancer Inst. 32, 507-517 (1964).
8. Sato, C., Int. J. Radiat. Biol., vol.18, n.5, 483-485 (1970).
9. Michalowski, A., Exp. Cell Res., 32, 609 (1963).
10. Sato, C., Sakka, M., Tohoku J. Exp. Med. 100, 375-381 (1970).
11. Castellani, A., Sedati, P., Belloni, P., Haematologica, 58, 915-930 (1973).
12. MacKinney, A.A., Stohlman, F., Brecher, G., Blood, 19, 349-358 (1962).
13. Peters, J.H., Hausen, P., Eur. J. Biochem. 19, 502-513 (1971).
14. Plagemann, P.G.W., Richey, D.P., Zylka, J.M., Erbe, J., Experimental Cell Research, 83, 303-310 (1974).

15. Wohlhueter, R.M., Marz, R., Plagemann, P.G.W., J. Membrane
    Biol. 42, 247-264 (1978).
16. Kwok, C.S., Chapman, I.V., Int. J. Radiat. Biol. 32, n.5,
    409-429 (1977).
17. Cleaver, J.E., Rad. Res. 37, 334-348 (1969).
18. Cleaver, J.E., Proc. of the Photobiology Meeting on "Factors
    Affecting Cellular Photosensitivity", Ed. A. Castellani,
    Annali Ist. Sup. San. 5, 360-366 (1969).
19. Ben-Hur, E., Ben-Ishai, R., Photochemistry and Photobiology,
    13, 337-345 (1971).
20. Francis, A.A., Dean Blevins, R., Carrier, W.L., Smith, D.P.,
    Regan, J.D., Biochimica et Biophysica Acta, 563, 385-392
    (1979).
21. Hanawalt, P.C., Research in Photobiology, Ed. A. Castellani,
    285-292, Plenum Press, New York (1977).
22. Castellani, A., Sedati, P., Biondi, G., "Radiobiologia del
    linfocita", Ed. L. Oliva, Atti dell'XI Congresso Nazionale
    dell'A.I.R.B.M., 39-51 (1975).
23. Sato, C., Kojima, K., Matsuzawa, T., Int. J. Radiat. Biol.
    20, n.1, (1971).
24. Sato, C., Kojima, K., Radiation Research, 60, 506-515 (1974).
25. Castellani, A., Biondi, G., Atti del XX Congresso Nazionale
    dell'A.I.F.S.P.R., 259-268, Bologna (1977).
26. Castellani, A., Biondi, G., Radiobiologia dei Tumori, Ed. C.
    Biagini e M. Di Paola, EMSI, 355-364, Rome (1977).
27. Lavin, M.F., Kidson, C., Nucleic Acids Research 4, n.11,
    4015-4022 (1977).

SCREENING FOR DEFICITS IN DNA REPAIR

USING HUMAN LYMPHOCYTES[#]

Shyam S. Agarwal* and Lawrence A. Loeb

The Joseph Gottstein Memorial Cancer Research Laboratory
University of Washington School of Medicine
Department of Pathology SM-30
Seattle, WA 98195

and

The Institute for Cancer Research
Fox Chase Medical Center
7701 Burholme Avenue
Philadelphia, PA 19111

INTRODUCTION

With the growing awareness concerning environmental carcinogenesis, it has become important to develop methods to quantitate
the response of individuals to carcinogens. With these tests in
hand, there will be at least three important distinct areas requiring extensive investigation: 1) the determination of which
chemicals in the environment initiate and promote malignant changes;
2) the delineation of groups of individuals having genetic differences in their response to particular carcinogens; and 3) the
development of methods to quantitate damage to cellular DNA which
has incurred from prior exposure to carcinogens. Tests to screen
for chemical carcinogens based on mutagenicity have been established in tissue culture, in bacteria, and in cell-free systems.
The definitive carcinogenicity of environmental agents must ulti-
mately rest on the results from animal test systems and from epidemiological data in human populations. For measuring genetic
variations in the response of individuals to carcinogens and possibly for analyzing accumulated damage to DNA, human peripheral
lymphocytes are a particularly advantageous biological system.

Lymphocytes obtained from peripheral blood are in the G_0 stage of the cell cycle. These non-dividing cells are unusually sensitive to gamma irradiation and perhaps to other agents that damage DNA (1). Phytohemagglutinin (PHA) and other mitogens stimulate these cells to undergo subsequent cell division (2). This stimulation involves gene activation and the synthesis of a number of key enzymes involved in DNA replication (3). DNA synthesis in PHA-stimulated lymphocytes starts approximately 20 to 24 hr following the addition of PHA, and the first mitosis can be observed by 40 hr. Our earlier studies have shown that the initiation of DNA synthesis is immediately preceded by the induction of DNA polymerase (3,4), and that the latter is dependent on prior RNA and protein synthesis (5). At the time of DNA replication, the increase of DNA polymerase activity primarily reflects an increase in the activity of DNA polymerase $-\alpha$ (6,7).

We reasoned that a highly sensitive index of functional DNA repair might be obtained by measuring the ability of cells to respond to phytohemagglutinin after exposure to agents that damage DNA (8). Our rationale assumes that the lack of adequate DNA repair prevents subsequent DNA replication and cell divisions. This rationale is similar to that underlying measurements of DNA repair in mammalian fibroblasts, on the basis of colony-forming ability after irradiation (9). Prior to PHA-stimulation, the small lymphocyte is unusually sensitive to X-irradiation; the D_0 for horse lymphocytes being 20 rad (1). Since quantitation is by DNA replication, the amount of thymidine incorporated in each cell is three to five orders of magnitude greater than that which would be observed during DNA repair. Thus, accurate measurements can be obtained on small numbers of lymphocytes.

RESULTS

Assay for Functional DNA Repair

An outline of the assay is diagrammed in Fig. 1. Cultures of human lymphocytes maintained in micro-well plates (1 X 10^5 cells/ 0.1 ml in each well) or in test tubes 1 X 10^6 cells/ml in each tube) are exposed to a DNA damaging agent (zero time). Four hours thereafter, a time at which the rapid phase of DNA repair has been completed, the cells are stimulated to undergo DNA division with phytohemagglutinin. After 96 hours, DNA polymerase activity and DNA replication are measured simultaneously by liquid scintillation spectroscopy. The measurement of [^{3}H]-thymidine incorporation is indicative of DNA replication in those cells that have successfully repaired their DNA damage. DNA polymerase activity is measured using lymphocyte lysates with added activated DNA as a template and all four nucleotides, one of which is labeled with [α-^{32}P]. DNA polymerase activity provides an independent criteria for cell survival.

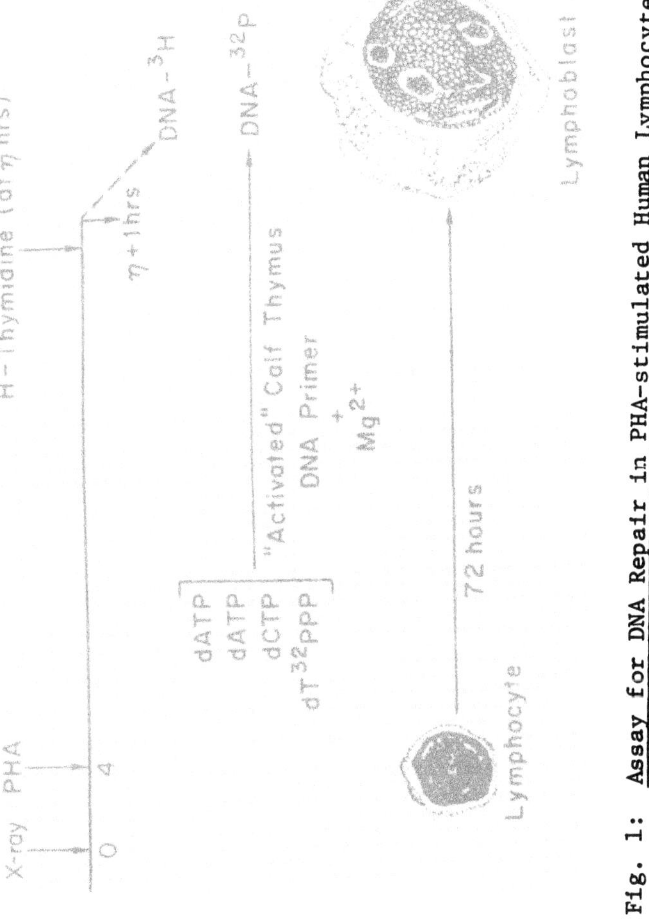

Fig. 1: Assay for DNA Repair in PHA-stimulated Human Lymphocytes

X-Ray Dose-Response Curve

The response of lymphocytes to PHA after exposure to different doses of X-irradiation is shown in Fig. 2. In this experiment, the cells were irradiated with X-ray doses between 25 and 2,400 rads. PHA was added to the cultures four hours after irradiation and the rate of DNA replication was measured at 82 hours. Irradiation of lymphocytes with up to 100 rads had no significant effect on the rate of thymidine incorporation. At doses between 100 and 800 rads, [3H] thymidine incorporation was diminished, and the decrease was proportional to the X-ray dose. In contrast, the induction of DNA polymerase was not decreased. If anything, the amount of DNA polymerase activity was greater than that observed in un-irradiated controls. Similar results were obtained when cultures were harvested at times between 72 and 96 hours after stimulation with PHA. Also, irradiation by itself did not result in any detectable increase in DNA polymerase activity or in thymidine incorporation when assayed at 92 hours. Thus, prior irradiation of lymphocytes diminishes PHA-stimulated DNA synthesis, and this decrease is proportional to the amount of X-irradiation.

Repair of X-Ray Damage in Normal Individuals

The effect of PHA-stimulation on irradiated lymphocytes is a fairly reproducible phenomenon. The degree of variation within replicate cultures is less than 10%, and repeated studies on lymphocytes from the same individual obtained at different times yield similar results. However, the response of lymphocytes from different individuals varies. The X-ray sensitivity of lymphocytes from 16 normal, healthy volunteers is shown in Figure 3. The lymphocytes from two of these individuals were relatively resistant to X-irradiation. This could be due to differences in the kinetics of the response to PHA. In the remaining 14, the rate of [3H] thymidine uptake in irradiated lymphocytes compared to unirradiated controls varied between 52 and 81% at 200 rads, 30 and 59% at 400 rads, and 18 and 37% at 800 rads. So far, we have not seen marked sensitivity to X-irradiation in lymphocytes from any of the normal, healthy individuals.

Kinetic Studies

Measurements have been made on the amount of DNA polymerase activity and the rate of thymidine incorporation in irradiated lymphocytes (400 rads) at different times after PHA-stimulation. The time of onset of the increased DNA polymerase activity after PHA-stimulation is essentially the same in un-irradiated cultures. In contrast, there is a delay in the initiation of DNA replication. After this initial delay, there is a rapid increase in DNA synthesis, so that at 42 hours the rate of DNA synthesis in irradiated

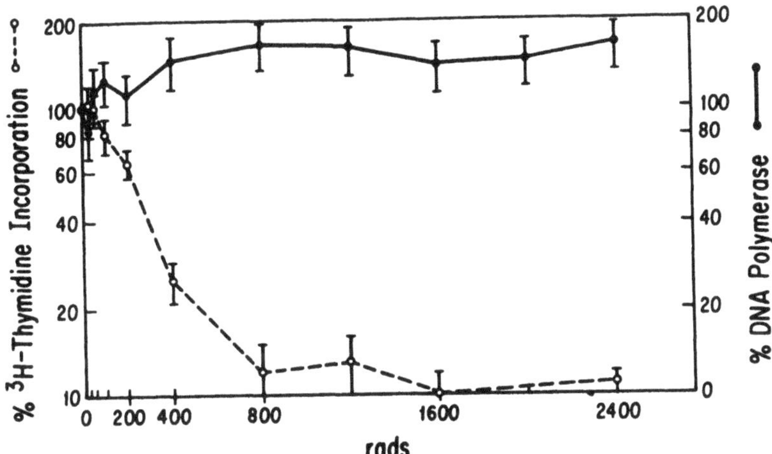

Fig. 2:   Effect of X-rays on Lymphocyte Stimulation

     Human peripheral lymphocytes were separated from a normal,
healthy donor on Ficoll-Hypaque gradients as described by Boyum (11).
They were suspended in RPMI-1640 medium supplemented with 2 mM
l-glutamine, 20% fetal calf serum, penicillin (100 units/ml), streptomycin (100 µg/ml) and 20 mM Hepes buffer. Aliquots of 1 X 10^6
cells/ml were distributed in 16 X 125 mm Falcon plastic culture
tubes. The cells were irradiated with 225 KVP X-rays from a
General Electric Maximar X-ray machine at a dose rate of 135 rads/
min at room temperature. After irradiation, cultures were returned
to humidified incubator at 37°C with 5% Co$_2$ in air. Four hours
after irradiation, lymphocytes were stimulated with PHA. At 68
hours after stimulation with PHA, the rate of [^{3}H]-thymidine incorporation (2.5 µCi/culture; 6.7 Ci/mmol) was measured for the terminal two hours in culture. The cell lysate of the same lymphocytes
was used for the measurement of DNA polymerase activity as described
previously (4). This assay measures both the DNA polymerase -α
and -β. Assays in the presence and absence of 10 mM n-ethylmaleimide (which specifically inhibits DNA polymerase -α) indicates
that both enzymes are unaffected by prior irradiation of lymphocytes (results not shown here). The rate of incorporation of [^{3}H]-
thymidine in unirradiated PHA-stimulated cultures (in quintupli-
cate) was 25,333 ± 3,129 cpm/2 hours and DNA polymerase activity
was 40.5 ± 6.3 pmol dTM^{32}P/hour. These values were taken to be
100%. The response of irradiated lymphocytes to PHA was computed
for each culture as percent of unirradiated control culture and
the mean ± S.D. of the resultant values are plotted.

cells and controls was similar. At 66 hours, after the addition of
PHA, thymidine incorporation of irradiated cells was only 31% of
the controls. This time corresponds to the second cycle of DNA
synthesis in PHA-stimulated lymphocytes. These results suggest
that a significant number of lymphocytes fail to undergo the second
round of replication and cell division. Inhibition of the DNA synthesis in irradiated lymphocytes after the first round of DNA replication is in accord with the concept of post-mitotic cell death.
However, it differs from true cell death in one important respect:
the DNA polymerase activity in irradiated lymphocytes continues to
increase. In un-irradiated lymphocytes, this increase has been
demonstrated to be dependent on continued RNA and protein synthesis (3 - 5). Thus, X-irradiation of lymphocytes selectively damages DNA and prevents subsequent DNA replication unless the damage
is repaired.

Studies in Human Disease

In order to determine whether the decrease in DNA replication
results from a lack of DNA repair, we have examined lymphocytes
from patients with known defects in DNA repair. The response of
lymphocytes from the patient with Xeroderma pigmentosum (XP) to
PHA after in vitro irradiation with UV is given in Table 1. The
patient was a 30-year-old female belonging to Group C of the XP repair defect classification (12). The lymphocytes from this patient
were markedly sensitive to UV irradiation. The rate of [^{3}H] thymidine incorporation in lymphocytes irradiated with 10 ergs/sq mm
was 8% of the unirradiated controls. In comparison, in lymphocytes
from the normal, healthy volunteer, the rate of [^{3}H] thymidine incorporation was 28% of the controls even after irradiation with an
8-fold higher dose (80 ergs/sq mm). In contrast, the value of thymidine incorporation in lymphocytes from the patient with XP after
exposure to X-ray fall within the lower limits of the normal distribution.

The response of lymphocytes from a patient with ataxia telangiectasia to PHA after in vitro X-irradiation is also given in
Table 1. Inhibition of the rate of thymidine incorporation in PHA-
stimulated irradiated lymphocytes from this patient was much
greater than that in the matched normal control at all doses. For
example, after irradiation with 100 rads, the rate of [^{3}H] thymidine incorporation in lymphocytes from the patient with ataxia telangiectasia was 49% compared to 91% in age and sex-matched simultaneous normal controls. With reference to a larger group, the response to PHA in irradiated lymphocytes from patients with ataxia
telangiectasia was lower than that of any normal, healthy individual studied so far (see Fig. 3). However, as in lymphocytes from
normal individuals, the induction of DNA polymerase was not inhibited.

Table 1

Response of Lymphocytes with known DNA Repair Deficits

DNA Damage	$[^3H]$ Thymidine Incorporation (Percent of Un-irradiated Cultures)	
UV (ergs/sq mm)	Patient with Xeroderma	Matched Control
10	8 ± 3.0	N.T.
40	1 ± 0.3	67 ± 16
80	1 ± 0.4	28 ± 5
160	N.T.	3 ± 1
X-ray (rads)	Patient with Ataxia	Matched Control
-50	75 ± 9	92 ± 5
100	49 ± 7	91 ± 6
200	28 ± 6	66 ± 4

Lymphocytes were cultured and exposed to either UV or X-irradiation as indicated. N.T. indicates Not Tested. The P value for the comparison of the patient with XP and control is 0.01. The P value for the patient with ataxia and control is 0.0005.

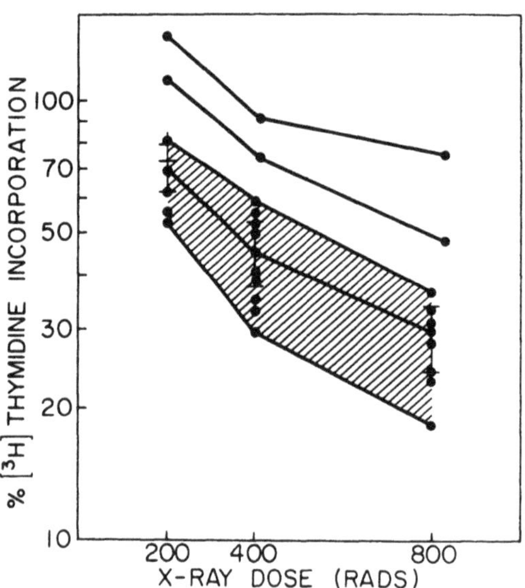

Fig. 3: <u>DNA Repair in Normal, Healthy Volunteers</u>

The shaded area represents the results of analyzing lymphocytes from 16 normal, healthy volunteers. The shaded area is the
range of values obtained for [3H]-thymidine incorporation. The
line connecting the mid-points is the mean values and the vertical
line is one standard deviation. The two upper-most lines are also
from normal individuals whose lymphocytes were markedly resistant
to X-irradiation. Taken from Agarwal and Loeb (8), with permis-
sion.

DISCUSSION

The results of this study show that the replicative response
of _in vitro_ irradiated lymphocytes to PHA can provide an indirect
measure of the ability of the cells to repair damage to DNA.   In
normal individuals, the inhibition of the rate of [^{3}H] thymidine
incorporation in irradiated lymphocytes upon stimulation with PHA
was found to parallel the radiation dose. With the use of this
assay, lymphocytes from patients with ataxia telangiectasia and XP
were found to be unusually sensitive to the effects of X-irradiation and UV irradiation, respectively.   These disease states are
known to have deficits in DNA repair by these agents (12, 13).
Presumably, if the DNA of the cell is not adequately repaired, DNA
replication is inhibited in subsequent generations.

The sensitivity of lymphocytes from the patient with XP was
much more marked to UV irradiation than to X-irradiation. This
differential defect in DNA repair was first observed by Cleaver (14)
in skin fibroblasts from these patients. Subsequent work has shown
that XP cells are also differentially sensitive to different chemicals (15). It should also be emphasized that patients with XP
exhibit a considerable degree of genetic heterogeneity. On the
basis of complementation studies, at least six distinct groups of
excision repair defects have been identified. The XP patient in-
cluded in this study belonged to Group C. The rate of DNA repair
in the cells from patients of this group has been reported to be
10 to 25% of normal (12). It would be interesting to study whether
the response of lymphocytes in the system reported here can identify individuals with more severe (groups A and B) or less severe
(group D) repair defects.

Irradiation of lymphocytes with X-rays has been used to block
the response of stimulator cells in one-way mixed lymphocyte culture (MLC) reactions (16). This use of X-rays was based on the
observation that X-irradiated lymphocytes do not incorporate thymidine upon PHA stimulation while their antigenic ability to stimulate the responder cells is maintained. However, our results
indicate that X-irradiated lymphocytes continue to remain metaboli-
cally active as is shown by the near-normal levels of DNA polymerase activity. They must synthesize RNA and protein and may
release mitogenic factors. This may account for the back-stimulation of cells in one-way MLC reactions where X-irradiation is used
for inhibiting the response of the stimulating cells (17). From a
practical point, much higher doses of irradiation (6400 rads) may
be required to eliminate this back-stimulation.

Our results indicate that the response of irradiated lymphocytes to PHA can be used to screen for deficits in DNA repair.
Peripheral lymphocytes can readily be obtained from large numbers

of individuals and long-term growth in culture is not required.
Other methods for measuring DNA repair are considered in other
chapters in this book, and include measurement of: (a) unscheduled
DNA synthesis; (b) excision of UV-induced thymidine dimers;
(c) non-semiconservative DNA synthesis (repair replication);
(d) post-replication repair; (e) rejoining of single-strand breaks
after X-irradiation; (f) host cell reactivation of irradiated
viruses; and (g) survival of fibroblasts by colony formation.
Except for the measurement of unscheduled DNA synthesis, most of
these methods are complex and time-consuming. Most importantly,
the method outlined in this paper should be applicable to measuring DNA repair after exposure to any DNA-damaging agent.

#This study was supported by grants from the National Institutes of
Health (CA-11524, CA-12818, CA-06551) and the National Science
Foundation (PCM 76-80439), by grants to the Institute for Cancer
Research from the National Institutes of Health (CA-06927, RR-05539),
and by an appropriation from the Commonwealth of Pennsylvania.

*Permanent Address: Department of Medicine, K.G. Medical College,
Lucknow University, Lucknow, India.

REFERENCES

1. Dewey, W.C. and Brannon, R.B., Int. J. Radiat. Biol., $\underline{2}$, Page
   229, 1976.

2. Nowell, P.C., Cancer Res., $\underline{20}$, Page 462, 1960.

3. Agarwal, S.S. and Loeb, L.A., Cancer Res., $\underline{32}$, Page 107, 1972.

4. Loeb, L.A., Agarwal, S.S. and Woodside, A.W., Proc. Natl.
   Acad. Sci. USA, $\underline{61}$, Page 827, 1968.

5. Loeb, L.A., Ewald, J.L. and Agarwal, S.S., Cancer Res. $\underline{30}$,
   Page 2514, 1970.

6. Meyer, R.J., Smith, R.G. and Gallo, R.C., Blood $\underline{46}$, Page 509,
   1975.

7. Bertazzoni, U., Steffanini, M., Noy, G.P., Giulotto, E.,
   Nuzzo, F., Falaschi, A. and Spadari, S., Proc. Natl. Acad.
   Sci. USA, $\underline{73}$, Page 785, 1976.

8. Agarwal, S.S., Brown, D.Q., Katz, E.J. and Loeb, L.A., Cancer
   Res. $\underline{37}$, Page 3594, 1977.

9. Taylor, A.M.R., Harnden, D.G. and Arlett, C.F., Nature, <u>258</u>, Page 427, 1975.

10. Loeb, L.A. and Agarwal, S.S., Exptl. Cell Res. <u>66</u>, Page 299, 1971.

11. Boyum, A., Scand. J. Clin. Lab. Invest., <u>21</u>, Page 31, 1968.

12. Robbins, J.H., Kraemer, K.H., Lutzmer, M.A., Festoff, B.W. and Coon, H.G., Ann. Intern. Med., <u>80</u>, Page 221, 1974.

13. Patterson, M.C., Smith, B.P., Lohman, P.H.M., Anderson, A.K. and Fishman, L., Nature, <u>260</u>, Page 444, 1976.

14. Cleaver, J.E., Nature, <u>218</u>, Page 652, 1968.

15. Stich, H.F., San, R.H.C., Miller, J.A. and Miller, E.C., Nature, <u>238</u>, Page 9, 1972.

16. Kasakura, S. and Lowenstein, L., Proc. Soc. Exp. Biol. Med., <u>125</u>, Page 355, 1967.

17. Thorsby, E., Transplant. Rev., <u>18</u>, Page 51, 1974.

DNA REPAIR IN SOME CANCER-PRONE CONDITIONS AND IN PATIENTS
WITH CUTANEOUS MALIGNANCIES

Bo Lambert and Ulrik Ringborg

Department of Clinical Genetics and Radiumhemmet

Karolinska Hospital, 104 01 Stockholm, Sweden

INTRODUCTION

Increasing evidence suggests that human cancer and mutation
may arise as a consequence of environmental exposure to chemical
and physical agents (1, 2). Many of these agents probably exert
their carcinogenic and mutagenic properties by a direct or indirect
interaction with the genetic material (3). A number of structural
modifications of the DNA can be overcome by cellular repair mechanisms (4). If these DNA repair mechanisms are defective, or if they
are not able to restore the original DNA molecule without causing
errors in the DNA sequence, mutations and neoplasms may result(5).
Thus, DNA repair mechanisms should be considered as important biological defence mechanisms against mutagenic and carcinogenic events.

Some human repair deficient diseases have been described, which
are all associated with an increased cancer risk (5, 6). These conditions have in common an inherited mutation which may be the cause
of or else indirectly related to the observed repair deficiency.
Cells from subjects with these diseases show in general, although
not invariably, decreased survival and enhanced mutation rates when
exposed to chemical or physical mutagens in vitro.

In analogy with these observations, one could speculate that
heterozygous carriers of genes for repair deficient disorders (7)
as well as otherwise healthy subjects who may have a slightly reduced DNA repair capacity in the lower range of the normal variation,
would also run an increased risk of acquiring cancer. Especially in
situations where carcinogenic exposure may be increased over normal,
e.g. in some industrial working conditions or excessive exposure to
sunlight, the identification of such presumptive risk individuals

should be of importance to medical health protection.

We have approached this problem by investigating DNA repair synthesis in leukocytes from a number of healthy subjects, as well as subjects with Down's syndrome and with different types of cutaneous malignancy, some of which have been causally related to sunlight exposure.

METHODS

There are several useful methods for the measurement of one or other parameter of DNA repair in mammalian cells (8). Hydroxyurea-insensitive incorporation of ^{3}H-thymidine into unstimulated human leukocytes can be used to estimate DNA repair synthesis (9, 10), which is one step in the pre-replicative excision repair mechanism. Since this technique is relatively simple, reproducible, and requires only a small amount of venous blood (10, 11), it is well suited for a survey of DNA repair capacity of human individuals.

Peripheral leukocytes are obtained from the buffy coat of 10 - 20 ml of freshly collected heparinized venous blood. The cells are washed in phosphate-buffered saline (PBS) and counted. Aliquots of the cell suspension are transferred to Petri dishes for UV-irradiation or to test tubes for treatment with chemical agents. After this treatment the cells are incubated at 37°C in Parker 199-medium for 30 min in the presence of 10^{-2}M hydroxyurea (HU). Methyl-^{3}H-thymidine is then added and incubation continued for 2 hrs. After three washes in trichloroacetic acid (TCA) the cells are resuspended in 70% ethanol and collected on glass-fiber filters. The non-TCA-extractable radioactivity on the filters is measured by liquid scintillation counting (for details of the technique, see 10, 11).

The preincubation with HU (see above) serves to depress the replicative DNA synthesis which is normally going on in a small fraction of peripheral leukocytes. It has been shown earlier (9, 10, 12) that HU does not apparently affect the DNA repair synthesis in this test system.

Representative results of HU-insensitive ^{3}H-thymidine-incorporation after treatment of human leukocytes with ultraviolet light (UV) and a number of cytostatic agents are shown in Fig. 1. It is clear that there is a dose-dependent increase of the incorporation after UV-irradiation as well as after treatment of the cells with the directly acting alkylating agents MMS, HN$_2$, melphalan and chlorambucil. The UV-induced incorporation reaches a plateau at higher

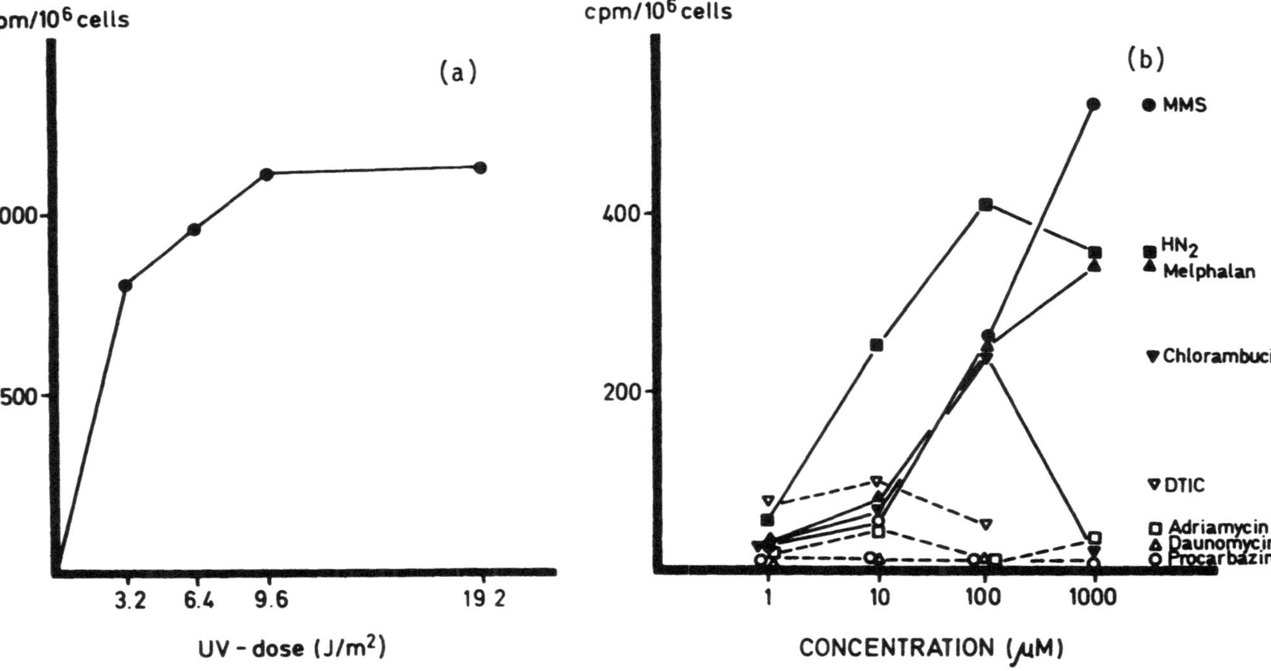

Fig. 1. DNA repair synthesis in human lymphocytes irradiated with UV-light (a) or treated with chemical agents (b) in vitro. The hydroxyurea-insensitive ^{3}H-thymidine incorporation (indicating DNA repair synthesis) is expressed in cpm per 10^6 cells after subtraction of the background activity in non-treated cells (60-80 cpm per 10^6 cells).

dose levels (Fig. 1a) probably indicating a saturation of the repair system. Such a plateau is not seen after treatment with the alkylating agents (Fig. 1b), probably because of the lethal effects of these chemicals at higher doses. No increase of ^{3}H-thymidine-incorporation is observed in cells treated with DTIC (dichloro-triazeno-imidazole-carboxamide), procarbazin, adriamycin and daunomycin (Fig. 1b). The two former drugs probably exert their cytostatic action by alkylation and degradation of DNA, but only after metabolic conversion in vivo. Adriamycin and daunomycin are antibiotics which intercalate in the DNA and inhibit effectively the replicative DNA synthesis. Although these drugs do not evoke DNA repair synthesis by themselves (Fig. 1b) they are effective inhibitors of DNA repair synthesis induced by UV, MMS and HN_2 (13).

These results show that HU-insensitive incorporation of ^{3}H-thymidine in human leukocytes can be used as a simple measure of DNA repair synthesis evoked by UV-irradiation as well as by directly acting alkylating agents.

RESULTS AND DISCUSSION

Subsequent studies have been devoted to analyses of the capacity for UV-induced DNA repair synthesis in various groups of subjects. For each individual a dose-response curve as in Fig. 1a has been established. The mean value for the ^{3}H-thymidine-incorporation (in cpm per 10^6 cells) at 9.6 and 19.2 J/m^2 (minus the background incorporation of non-irradiated cells) has been used as a measure of the individual DNA repair capacity.

The methodological variation was estimated by making 10 separate measurements on one subject. The standard deviation (S.D.) was 12.4% of the mean value which is considerably less than the S.D. of 39.5% obtained from measurements of 48 other subjects. This difference is statistically significant (14) and indicates a variation in the capacity for UV-induced repair synthesis between individuals.

Part of the individual variation is probably an effect of age, since there is a significant, negative correlation between DNA repair synthesis and increasing age (Fig. 2). The average DNA repair capacity decreases by about 25% between 20 and 90 years of age and, as indicated in Table 1, subjects over 60 years show a significantly lower DNA repair synthesis than subjects below 60 years of age (14).

Down's syndrome (DS) is associated with an increased incidence of leukaemia (15). There is an increased frequency of chromosome aberrations after treatment with X-ray (10) and chemical mutagens (16) in DS cells compared to normal cells. These observations suggest that DS may involve some disturbance of DNA repair mechanisms. As

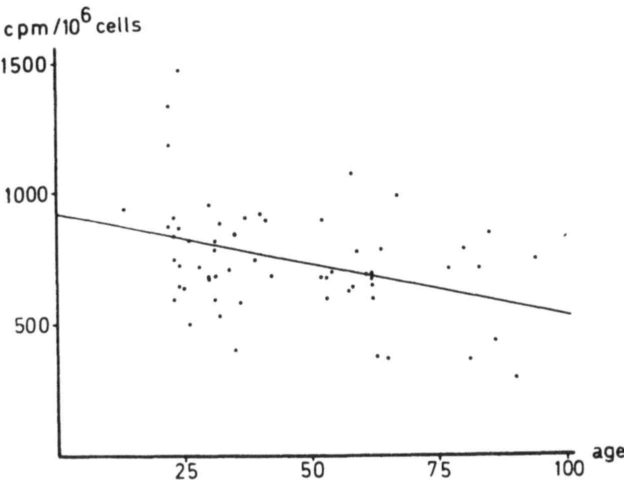

Fig. 2. Correlation plot of UV-induced DNA repair synthesis in lymphocytes against age of the donor. The straight line is fitted by regression analysis (r=-0.38, n=58, P<0.005).

Table 1.UV-induced DNA repair synthesis in peripheral lymphocytes from control subjects and patients with various disorders.

Clinical diagnosis	No. of subjects	DNA repair synthesis (%of controls)[*]	Signi- ficance (P<)
Controls 13 - 59 yrs	30	100	
60 - 94 yrs	18	72.6	0.01
Down's syndrome	10	70.4	0.01
Actinic keratosis	10	70.4	0.01
Basal cell naevus syndrome	4	77.1	0.05
Malignant melanoma	21	95.3	NS
Basal cell carcinoma	6	106.2	NS
Epidermodysplasia verruciformis	1	60.7	—
Chronic lymphatic leukemia	10	87.1	NS

[*]Subjects of the various groups of disorders were compared to controls of similar age.

shown in Table 1 and more detailed previously (10), subjects with
DS show only about 70% of the normal UV-induced DNA repair synthesis.

Many observations indicate that solar UV-radiation has carcinogenic activity on dermal cells (17), and sunlight is considered
as a possible etiological factor in e.g. actinic keratosis, basal
cell carcinoma and malignant melanoma (17-19). It was therefore of
interest to study UV-induced DNA repair synthesis in subjects with
these disorders (11, 20, 21). As shown in Table 1, patients with
melanoma and multiple basal cell carcinoma show DNA repair synthesis
within the normal range (20, 21). Malignant melanomas are of different histology. There may be a difference in the connection between
UV-exposure and superficial spreading melanomas compared to nodular
melanomas. However, patients with superficial spreading melanoma
also show a normal UV-induced DNA repair synthesis (21).

Actinic keratosis is a premalignant lesion of the epidermis,
which, if left untreated, will slowly progress to squamous cell
carcinoma in 12 - 13% of cases. A strong relationship between this
disorder and sun exposure has been demonstrated by epidemiological
methods (19). The results shown in Table 1 indicate that subjects
who develop actinic keratosis may have a lower capacity for UV-induced DNA repair synthesis than healthy subjects of similar age (11).
This observation was recently confirmed in a study of unscheduled
DNA synthesis in fibroblasts derived from patients with actinic
keratosis (22).

Basal cell naevus syndrome is an autosomal dominant disorder
with multiple basal cell carcinomas, odontogenic cysts and skele-
tal abnormalities. The carcinomas appear when the patients are very
young, and often the first localization is in the face. Very likely,
solar radiation is a factor contributing to the malignant transformation. Four subjects with this disorder have been studied so far.
As shown in Table 1 the average UV-induced DNA repair synthesis in
these patients is about 20% lower than that in healthy control subjects of similar age (23). These results suggest that a reduced DNA
repair capacity together with other host-factor(s) may predispose
to malignant transformation in this disorder.

Epidermodysplasia verruciformis(EV) is a rare skin disease in
which plane papules appear in teenage. Common wart virus has been
identified in such lesions. Skin malignancy is common in EV and the
viral infection has been suggested as an oncogenic factor. One patient with this disease was subjected to repeated analysis of UV-induced DNA-repair synthesis (24). The patient had multiple manifesta-
tions of actinic keratosis, basal cell carcinoma, Bowen's disease
and squamous cell carcinoma, and the tumors first appeared on sun
exposed areas. The UV-induced DNA repair synthesis was found to be
low compared to control subjects of similar age (24, Table 1).

This result suggests that in EV the predisposition to acquire a
possibly oncogenic virus infection in combination with environmental influences (e.g. sun-light) and a decreased capacity for
UV-induced DNA repair synthesis may enhance somatic mutations and
malignant transformation.

Studies of DNA repair synthesis in lymphocytes from patients
with chronic lymphatic leukaemia (CLL) have given conflicting results. Frey-Wettstein and collaborators (25) found no difference
in UV-induced unscheduled DNA synthesis in lymphocytes from healthy
donors and CLL patients, while Huang and coworkers (26) reported
an increased DNA repair synthesis in UV-irradiated CLL-lymphocytes
compared to normal lymphocytes. Our results (27) agree with those
of Frey-Wettstein et al. Ten patients with CLL have been studied.
As indicated in Table 1 there is no statistical difference between
normal and CLL-lymphocytes in the UV-stimulated, HU-insensitive
^{3}H-thymidine-incorporation (27). The results of Huang and coworkers
(26) may be explained by the omittance of hydroxyurea in their incubation medium, which leads to a spurious calculation of DNA repair
synthesis, because the influence of UV-irradiation on DNA synthesis
in replication cells is not taken into account (27).

Our results indicate the necessity of eliminating the replicative DNA synthesis in order to obtain an adequate determination of
the repair synthesis, even though the amount of replicative DNA
synthesis is low in peripheral lymphocytes. This can be achieved by
including hydroxyurea in the incubation medium, which depresses DNA
replication to a low and reproducible background level in all conditions so far studied (10, 11, 13, 14, 20, 21, 23, 27), and permits
adequate calculations of UV-induced DNA repair synthesis.

SUMMARY

DNA repair synthesis was estimated in human peripheral lymphocytes in vitro by measurements of the hydroxyurea-insensitive incorporation of ^{3}H-thymidine after treatment of the cells with UV-
light and chemical mutagens.

UV-irradiation, methyl methane sulphonate and the directly
acting antineoplastic drugs nitrogen mustard, melphalan and chlorambucil caused a dose-related increase in DNA repair synthesis. No
repair stimulation was observed after cellular treatment with the
indirect agents dichloro-triazeno-imidazole-carboxamide (DTIC) and
procarbazine, which require metabolic conversion in vitro to exert
cytostatic effects, nor with the DNA-intercalators adriamycin and
daunomycin.

The UV-induced DNA repair synthesis was studied in healthy

subjects of various ages, subjects with Down's syndrome and patients with different types of cutaneous malignancy in which UV-exposure has been incriminated as a causative factor. A significant decrease in repair synthesis compared to controls was observed in healthy control subjects at higher ages, in subjects with Down's syndrome, in patients with actinic keratosis and basal cell naevus syndrome. A single patient suffering from the rare dermatosis Epidermodysplasia verruciformis and multiple cutaneous malignancies, also showed a very low UV-induced DNA repair synthesis. No significant difference from the normal repair synthesis was observed in patients with malignant melanoma, basal cell carcinoma and chronic lymphatic leukaemia.

ACKNOWLEDGEMENT

The present work was supported by grants from King Gustaf Vth Jubilee Fund, the Swedish Medical Research Council, the Swedish Work Environmental Fund and Karolinska Institutet.

REFERENCES

1. Doll, R., Strategy for detection of cancer hazards to man, Nature 265:589-596 (1977).

2. Committee 17 of the Environmental Mutagen Society, Environmental Mutagenic Hazards, Science 187:503-514 (1975).

3. Miller, E.C., Some current perspectives on chemical carcinogenesis in humans and experimental animals, Cancer Res.38: 1479-1496 (1978).

4. Reagan, J.D. and Setlow, R.B., Repair of chemical damage to human DNA, in"Chemical mutagens, principles and methods for their detection," Vol. 3:151-169, A. Hollaender (ed.), Plenum Press, New York, London (1973).

5. Setlow, R.B., Repair deficient human disorders and cancer, Nature 271:713-717 (1978).

6. Cleaver, J.E., DNA repair processes and their impairment in some human diseases, in "Progress in genetic toxicology," 29-42, D. Scott, B.A. Bridges and F.H. Sobels (eds.), Elsevier, Amsterdam (1977).

7. Marx, J.L., DNA repair: New clues to carcinogenesis, Science 200:518-521 (1978).

8.   Cleaver, J.E., Methods for studying excision repair of DNA
     damaged by physical and chemical mutagens, in "Handbook of
     mutagenicity test procedures," 19-48, B.J. Kilbey, M. Legator,
     W. Nichols and C. Ramel (eds.), Elsevier, Amsterdam (1977).

9.   Evans, R.G. and Norman, A., Radiation stimulated incorporation
     of thymidine into the DNA of human lymphocytes, Nature 217:
     455-456 (1968).

10. Lambert, B., Hansson, K., Bui, T.H., Funes-Cravioto, F.,
     Lindsten, J., Holmberg, M. and Strausmanis, R., DNA repair
     and frequency of X-ray and UV-light induced chromosomeaber-
     rations in leukocytes from patients with Down's syndrome,
     Ann. Hum. Genet. 39:293-303 (1976).

11. Lambert, B., Ringborg, U. and Swanbeck, G., Ultraviolet in-
     duced DNA repair synthesis in lymphocytes from patients with
     actinic keratosis, J. invest. dermatol. 67:594-598 (1976).

12. Cleaver, J.E., Repair replication of mammalian cell DNA: Effects
     of compounds that inhibit DNA synthesis or dark repair,
     Radiation Res., 37:334-348 (1969).

13. Ringborg, U. and Lambert, B., Inhibition of DNA repair synthesis
     by some intercalating agents, in "DNA repair and late effects,"
     Proceedings of a symposium, Tel Aviv (1978) (in press).

14. Ringborg, U., Lambert, B. and Swanbeck, G., DNA repair in con-
     ditions associated with malignancy: Aging and actinic keratosis,
     in "Prevention and Detection of Cancer," 39-52, H.E. Nieburgs
     (ed.), Marcel Dekker, New York (1978).

15. Jackson, E.W., Turner, J.H., Klauber, M.R. and Norris, F.D.,
     Down's syndrome. Variation of leukaemia occurrence in institu-
     tionalized populations, J. Chron. Dis. 21:247-253 (1978).

16. Schuler, D., Dobos, M., Fekete, G., Hachay, T. and Nemeskéri, A.,
     Down's syndrome and malignancy, Acta Ped. Acad. Scient. Hung.
     13:245-252 (1972)

17. Urbach, F., Epstein, J.H. and Forbes, P.D., Ultraviolet car-
     cinogenesis: experimental, global and genetic aspects, in
     "Sunlight and Man," 259-283, M.A. Pathak, M.A. Harber, L.C. Segi
     and A. Kukita (eds.), University of Tokyo Press (1974).

18. Beardmore, G.L., The epidemiology of malignant melanoma in
     Australia, in "Melanoma and skin cancer," 39-64, W.H. McCarthy
     (ed.), Government printer, Sydney (1972).

19. Daniels, F., Sunlight, in "Cancer epidemiology and prevention,"
    126-152, D. Schottenfeld (ed.), Charles C. Thomas, Springfied,
    Ill. (1974).

20. Ringborg, U. and Lambert, B., Normal UV-induced DNA repair syn-
    thesis in peripheral leukocytes from patients with multiple
    basal cell carcinomas (in manuscript).

21. Ringborg, U. and Lambert, B., Normal UV-induced DNA repair syn-
    thesis in peripheral leukocytes from patients with malignant
    melanoma (in manuscript).

22. Sbano, E., Andreassi, L., Fimiani, M., Valentino, A. and
    Baiocchi, R., DNA-repair after UV-irradiation in skin fibro-
    blasts from patients with actinic keratosis, Arch. Derm. Res.
    262:55-61 (1978).

23. Ringborg, U., Lambert, B. and Swanbeck, G., Decreased UV-in-
    duced DNA repair synthesis in peripheral leukocytes from pati-
    ents with Basal Cell Naevus Syndrome, (in manuscript).

24. Hammar, H., Hammar, L., Lambert, B. and Ringborg, U., A case
    report including EM and DNA repair investigations in a derma-
    tosis associated with multiple skin cancers: Epidermodysplasia
    Verruciformis, Acta Med. Scand. 200:441-446 (1976).

25. Frey-Wettstein, M., Longmire, R. and Craddock, C.G., Deoxyribo-
    nucleic acid (DNA) repair replication of ultraviolet (UV) radi-
    ated normal and leukemic leukocytes, J. Lab. Clin. Med. 74:109-
    118 (1969).

26. Huang, A.T., Kremer, W.B., Laszlo, J. and Setlow, R.B., DNA re-
    pair in human leukaemic lymphocytes, Nature New Biol. 240:114-
    116 (1972).

27. Ringborg, U. and Lambert, B., Ultraviolet-induced DNA repair
    synthesis in leukocytes from patients with chronic lymphatic
    leukemia, Cancer Letters 3:77-81 (1977).

SISTER CHROMATID EXCHANGE IN HUMAN LYMPHOCYTES AS AN INDICATOR OF

DNA DAMAGE AND REPAIR IN VIVO

Bo Lambert

Department of Clinical Genetics, Karolinska Hospital

104 01 Stockholm, Sweden

INTRODUCTION

The development of staining techniques for the demonstration
of sister chromatid exchanges (SCE) (1) has introduced a new, promising tool in the evaluation of genotoxic effects of mutagenic and
carcinogenic agents. SCEs arise when the two chromatids within a
single chromosome exchange segments (Fig. 1). This usually occurs
at homologous loci, so the SCE itself may not alter the genetic information of the cell or its daughter cells. In fact, SCEs may represent a natural event of mitotic recombination, since present
evidence suggests that SCE occurs at a very low frequency in normal
somatic cells (2).

The formation of an SCE involves breakage of the double stranded DNA in both chromatids followed by an exchange of whole DNA du-
plexes. These events must occur during S-phase, since no SCEs are
seen in cells that do not undergo DNA-replication before observation at mitosis (3, 4). Although the nature of the lesions leading
to SCE is not fully understood it is very likely that SCEs arise as
a consequence of a repair process which permits DNA-replication to
continue beyond DNA lesions in the replication fork (5).

A great number of mutagens and carcinogens increase the frequency of SCE in animal cells in vitro (6-10) and in vivo (11-13) at
concentrations which are far lower than those required to produce
visible chromosome aberrations. It is clear, therefore, that SCE is
a very sensitive indicator of DNA-damage caused by these agents.
Using four mutagenic agents Carrano et al. (14) recently showed a
strong linear relationship between the mutation rate at the HGPRT-

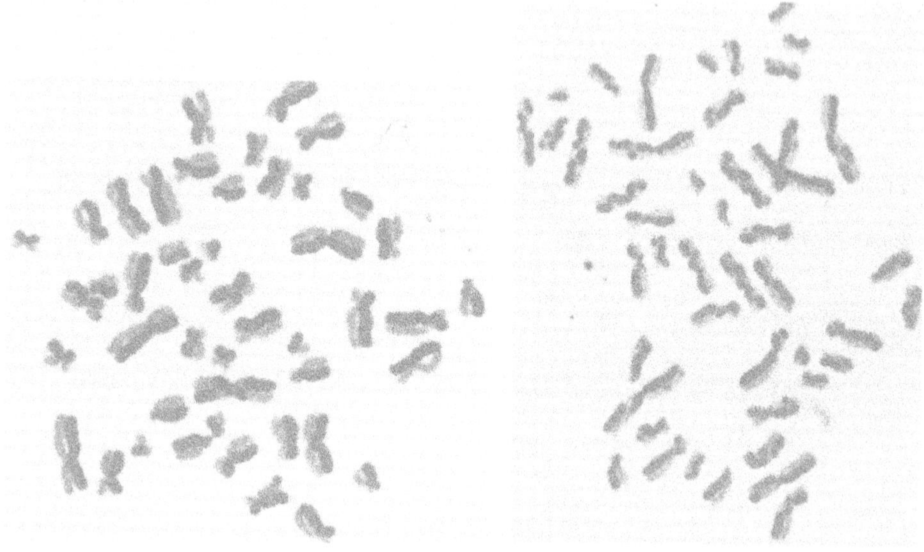

Fig. 1. Human lymphocyte chromosomes demonstrating SCE. Left: control
subject 15 SCEs. Right: melanoma patient 70 SCEs after CCNU-chemo-
therapy.

locus (producing resistance to 8-azaguanine) and the overall SCE-
frequency in CHO-cells. Thus, SCE may not only be a sensitive indicator of repairable DNA-damage, but also a useful indicator of
those specific and probably rare DNA lesions that are responsible
for mutagenesis and possibly carcinogenesis.

In order to evaluate the SCE-technique as an indicator of DNA-
damage in human cells after in vivo exposure to genotoxic agents,
we have determined SCE-frequencies in peripheral lymphocyte cultures
from 175 individuals, including children and adult control subjects,
occupationally exposed subjects and patients under cytostatic treatment for psoriasis and various types of cancer.

METHODS

A simple modification of the fluorescence plus Giemsa technique
described by Perry and Wolff (15) was used. Cultivation of phytohemagglutinin stimulated lymphocytes, staining procedure and SCE
counting were as described previously (16).

Bromodeoxyuridine (BrdU), which is needed to stain the chromatids differentially, was added at the start of the cultures at a
final concentration of 100 µM. The average number of SCEs per cell

per individual ("individual SCE-frequency") was based on counting
of at least 20 cells. In some of the cancer patients under study
the number of cells available for analysis did not amount to 20,
probably because of growth inhibition due to the chemotherapy.

RESULTS AND DISCUSSION

Control Subjects

Control subjects were studied in order to obtain information
about the variation of the SCE-frequency in the occupationally non-
exposed population, to trace the background factors which may influence the individual SCE-frequency, and to ascertain what level of
increase over the normal SCE-frequency may be considered significant.

Subjects used as controls were students drafted for military
service, medical students and blood donors of various professions.
They were all questioned about their working and social conditions,
medical treatment, hobbies, and private habits including smoking and
alcohol consumption. So far no other factor than smoking habits has
been found to influence the SCE-frequencies in this control group
(17). Fig. 2 shows the distribution of SCEs in a total of 1040 cells.
Among non-smokers there is a small deviation from the normal distribution in that cells with higher numbers of SCEs are slightly over-
represented. In heavy smokers there is a shift towards higher SCE-
numbers per cell. Only one out of 835 cells from non-smokers had
more than 30 SCEs (0.11%), while 10 out of 388 (2.57%) cells from
heavy smokers had 30 SCEs or more. Thus, the number of SCEs/cell in
the control population show a wide range of variation, from 3 to 49,
and cells with higher numbers of SCEs are found more frequently among
smokers (Fig. 2).

Table 1 shows compiled data of individual SCE-frequencies (i.e.
the average no of SCEs per cell per individual) for various groups
of subjects. It is obvious that there is a considerable heterogeneity between individuals in the adult control group. The individual
SCE-frequencies range from 7.0 to 24.2 SCEs/cell/individual. Again
the highest individual SCE-frequencies are found among smokers.
Children (age 0.2-7 years) have an SCE-frequency which is significantly lower than that of non-smoking adults (t-test, P<0.001).

These data on SCE-frequencies in control subjects show that
there is a considerable variation between subjects, which partly may
depend on individual differences in BrdU uptake (16). Another cause
of variation is smoking habits. As shown in Table 2 there is a dose-
dependent increase of SCE among smokers (17). Thus, when groups of
occupationally exposed subjects are compared with control subjects,

Table 1. SCE in Different Groups of Subjects

| Group of Subjects | Mean Number of SCE per Cell per Individual | | | | | |
| | Non-Smokers | | | Smokers | | |
	Number of Subjects	Group Mean±S.D.	Range	Number of Subjects	Group Mean±S.D.	Range
Controls						
children	12	9.2±2.1	7.0-13.8	-	-	-
adults	45	14.0±3.0	7.8-20.2	24	15.6± 3.3	9.2-24.2
Laboratory Workers						
mixed exposure	17	19.9±4.6	12.1-32.9	18	18.4± 3.5	14.4-25.3
ethylene oxide exposed	2	16.1±5.7	12.0-20.1	3	21.1± 6.9	16.0-28.9
Psoriasis Patients						
before treatment	8	13.7±3.1	10.4-20.0	10	16.8± 2.4	13.8-22.0
after PUVA-treatment	7	13.5±5.1	8.7-22.6	7	15.0± 3.8	10.4-22.1
Cancer Patients						
CCNU-therapy	5	35.8±7.7	25.5-44.5	5	33.3±10.5	23.3-48.1
other types of chemotherapy	7	15.2±5.8	6.0-24.8	5	17.2± 4.5	11.2-22.4

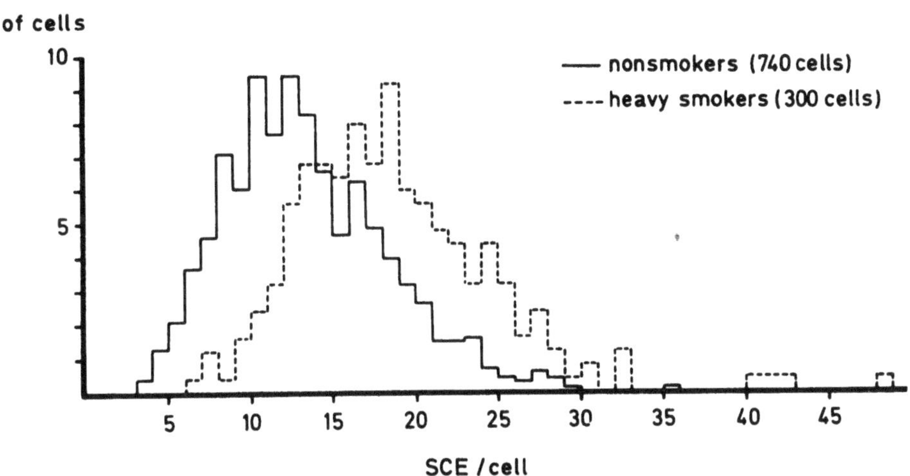

Fig. 2. Distribution of SCEs in cells from 27 non-smokers (————)
and 15 heavy smokers (-----).

Table 2. SCE in Smokers and Non-Smokers

Subjects		Mean No of SCE per Cell per Individual	
Groups [a]	No	Group Mean±S.D.	Range
non-smokers	65	14.0±3.0	7.8-20.6
moderate smokers	22	16.2±5.4	9.2-33.2
heavy smokers	22	17.6±2.7	12.8-24.2

a; moderate smokers were subjects taking 1 - 10 cigarettes
per day and heavy smokers >10 cigarettes per day.

smoking habits have to be taken into account. In agreement with
others (18, 19) we have not found age or sex to influence the SCE-
frequency in the adult population.

SCE-Frequencies in Patients Receiving

Cancer-Chemotherapy and PUVA-Treatment

Patients treated with potentially genotoxic agents in cancer-
chemotherapy or PUVA-therapy against psoriasis have been studied as
positive controls. The advantage with this approach is that the conditions of exposure can be controlled, permitting dose-effect and
time-effect relationships to be explored in greater detail.

PUVA-treatment, i.e. orally administered 8-methoxypsoralen
(8-MOP) followed by irradiation of the skin by longwave ultraviolet
light (UVA, 360 nm) 2-2.5 hr later is a recently introduced and effective therapy against severe psoriasis (20, 21). Under irradiation
with UVA, 8-MOP photoreacts with DNA producing monofunctional adducts to pyrimidine bases and interstrand cross-links (see 22). The
resulting inhibition of cell proliferation is likely to be one reason for the beneficial clinical effect on the disease. PUVA-treatment of human lymphocytes in vitro causes a significant, dose-related increase in chromosome aberrations (21) and sister chromatid exchange (23). Animal studies have shown that UVA-irradiation in combination with topical 8-MOP administration gives rise to an increased frequency of skin tumours (see 22). Thus, there is a potential
risk for mutagenic and/or carcinogenic effects of PUVA-treatment.

We have studied the frequency of SCE in psoriasis patients before and after 3-30 weeks of intensive treatment. The SCE-frequencies for all patients were within the control ranges and no treatment related changes were found (23). Since PUVA-treatment of human
lymphocytes in vitro gives rise to an increased frequency of SCE the
non-appearance of such a result in vivo could be due to insufficient
serum concentrations of 8-MOP and/or to a minimal penetration of
UVA-light to the circulating blood cells. These former alternatives
were investigated by taking blood from patients who had ingested the
standard dose of 8-MOP (0.6 mg/kg of body weight) two hours before
blood sampling and irradiating the blood in vitro with UVA in a dose
range similar to that given to the patient in a single clinical treatment. The results of this experiment (Fig. 3) show that the concentration of 8-MOP in the serum of the patients is high enough to
produce an increase in the frequency of SCE in the lymphocytes, if
only the cells receive a sufficient dose of UVA light. The probable
reason for the failure to detect an increased SCE-frequency after
sustained in vivo-treatment then is that the circulating lymphocytes
receive only a very small fraction of the total body dose of UVA
given to the patients (see table 1).

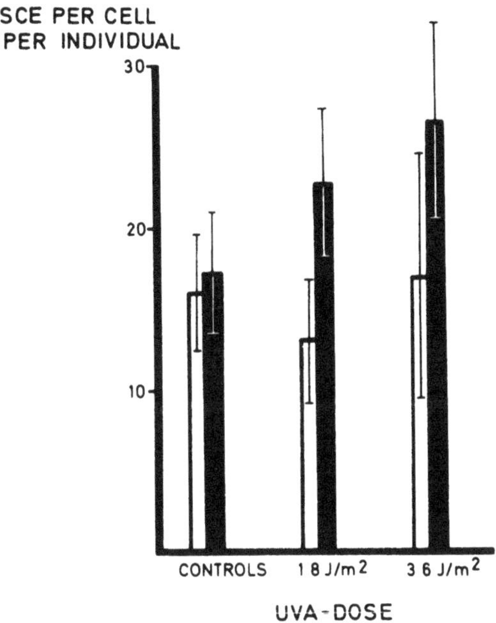

Fig. 3. Frequency of SCE in UVA-irradiated lymphocytes from psoriasis patients 2 hours after oral administration of 8-methoxypsoralen (8-MOP) (dark columns) and control patients receiving no 8-MOP (open columns). The bars indicate 1 S.D.

Cancer patients receiving chemotherapy have been studied in order to obtain information about which types of chemotherapeutic agents may promote SCE-formation in vivo, and to analyse the longevity of cellular damage giving rise to SCE. So far the most interesting observations have been obtained in patients receiving CCNU (lomustin) (24, 25). The SCE-analyses were usually made at mid-intervals between courses of treatment or immediately before the following course. Thus, time between treatment and SCE-analysis differs among the patients from about 2-8 weeks. All patients receiving CCNU as the single type of therapy or in combination with other drugs had at least a two-fold increase in the SCE-frequency compared to average control values (Table 1). Very high SCE-numbers were recorded in some cells from these patients (Fig. 1).

These results show that CCNU induces very long-lived DNA-lesions capable of producing SCE up to two months after drug administration. In contrast, patients treated with several other agents (e.g. bleomycin, 5-fluoracil, methotrexate, prednimustine, adriamycin, DTIC (dacarbazine), cyclophosphamide and actinomycin D) showed normal

SCE-frequencies 2-8 weeks after the administration (24, 25). It remains to be shown whether the long-lasting, SCE-promoting effect of CCNU treatment may be of particular significance for mutagenesis and the development of secondary neoplasms.

SCE-Analysis of Occupationally Exposed Subjects:

Laboratory Workers and Subjects Exposed to Ethylene Oxide

In previous works we have reported on the SCE-frequencies in personnel from a number of laboratories performing hormone analysis and organic chemical research (26, 27). One common denominator for these subjects is that they all have regular contact with organic solvents in their work, e.g. chloroform, toluene, diethyl ether, cyclohexane, xylene and methanol. Benzene was used in small quantities in some of the laboratories, in addition to a number of other chemicals and isotopes used in laboratory routine work. No one particular chemical, nor radioactivity, was used in excess quantities or handled in undue manners. Air concentrations of a number of organic solvents were, with few exceptions, well below threshold limit values. The average individual SCE-frequency in this group of laboratory workers was found to be significantly increased compared to current control values (26, 27). Later, more laboratory workers have been investigated and compared to the extended control material according to smoking habits (17). As shown in Table 1 there is a significant increase of SCE in non-smoking laboratory workers over that of non-smoking control subjects. However, no such difference was found between smoking control subjects and laboratory workers. In fact, no effect of smoking could be seen within the group of laboratory workers. This latter finding suggests that there is not a simple additive effect of different types of genotoxic exposure on the SCE-frequency.

Even if the specific cause of the increased SCE-frequency among non-smoking laboratory workers is not known, it seems reasonable to suspect some factor(s) in the working environment, e.g. organic solvents. Obviously, further studies are necessary to sort out the most potent SCE-inducing agent(s) in the laboratory environment. One way to do this would be to study personnel in laboratories or factories using very few chemicals, e.g. sterilization plants using ethylene oxide (EO). The potential mutagenicity and carcinogenicity of EO have been discussed by Ehrenberg and coworkers (28, 29). In order to investigate possible genotoxic effects of EO-exposure in vivo in man, we have studied the SCE-frequency in peripheral lymphocytes of five female workers occupied in a sterilization plant. The rationale for carrying out this study on such a small number of persons is the existence of relatively reliable, individual exposure data (29). The average SCE-frequency in the EO-exposed workers is 19.1±6.3, which is very similar to the average SCE-frequency in the laboratory workers (see Table 1). It is possible that the increased SCE-frequencies in the EO-exposed group is unspecifically related to the general laboratory environment. However, there is a positive, although not statistically significant, correlation (r = 0.7) between exposure dose of EO and an increase of SCE over the control level, which suggests a more specific effect of EO as well (30). We therefore consider these results as a primary indication of the genetic toxicity of occupational EO-exposure.

SUMMARY

The frequency of sister chromatid exchanges (SCE) was studied in cultures of human lymphocytes obtained from healthy control subjects, occupationally exposed individuals and patients under medical treatment with cytostatic agents.

The average SCE-frequency in non-smoking adult control subjects was 14.0 SCE/cell, in cigarette smokers 15.6 SCE/cell and in childrei (age 0.2-7 yrs) 9.2 SCE/cell. The effect of smoking habits on the SCE-frequency was further substantiated by the observation of a dose-related increase of SCE in moderate and heavy smokers.

The average SCE-frequency in non-smoking laboratory workers (19.9 SCE/cell) was significantly increased compared to non-smoking controls.

Psoriasis patients under sustained PUVA-therapy (8-methoxypsoralen) plus UVA-irradiation had SCE-frequencies within the control range. When blood from 8-methoxypsoralen treated patients was irradiated in vitro with increasing doses of UVA-light, a dose-related increase of SCE was demonstrated.

Cancer patients treated with 1-(2-chloroethyl)-3-cyclohexyl-1-nitrosourea (CCNU) were found to have increased SCE-frequencies up to 8 weeks after a single administration of the drug. No increase of SCE at this late time after chemotherapy was observed in cancer patients treated with a number of other agents.

ACKNOWLEDGEMENT

The present work was supported by grants from the Swedish Work Environmental Fund and Karolinska Institutet.

REFERENCES

1. Latt, S.A., Microfluorimetric detection of DNA synthesis in human chromosomes, _Proc. Nat. Acad. Sci._70:3395 (1973).

2. Kato, H., Spontaneous sister chromatid exchanges detected by a BUdR-labelling method, _Nature_ 251:70 (1974).

3. Wolff, S., Bodycote, J. and Painter, R.B., Sister chromatid exchanges induced in Chinese hamster cells by UV-irradiation of different stages of the cell cycle: the necessity for cells to pass through S, _Mutation Res._ 25:73 (1974).

4. Wolff, S., Chromosomal effects of mutagenic carcinogens and the nature of the lesions leading to sister chromatid exchange, _in_ "Mutagen induced chromosome damage in man," H.J. Evans and D.C. Lloyd (eds.), Edinburgh University Press, Edinburgh (1978).

5. Evans, H.J., Molecular mechanisms in the induction of chromosome aberrations, _in_ "Progress in genetic toxicology," B.A. Bridges, D. Scott and F.H. Sobels (eds.), Elsevier North Holland, Amsterdam (1977).

6. Perry, P. and Evans, H.J., Cytological detection of mutagen-carcinogen exposure by sister chromatid exchange, _Nature_ 258: 121 (1975).

7. Solomon, E. and Bobrow, M., Sister chromatid exchanges - a sensitive assay of agents damaging human chromosomes, _Mutation Res._, 30:273 (1975).

8. Popescu, N.C., Turnbull, D. and di Paolo, J.A., Sister chromatid exchange and chromosome aberration analysis with the use of several carcinogens and non-carcinogens, _J. Natl. Cancer Inst._ 59:289 (1977).

9. Abe, S. and Sasaki, M., Chromosome aberrations and sister chromatid exchanges in Chinese hamster cells exposed to various chemicals, _J. Natl. Cancer Inst._, 58:1635 (1977).

10. Takehisa, S. and Wolff, S., Induction of sister chromatid exchanges in Chinese hamster cells by carcinogenic mutagens requiring metabolic activation, _Mutation Res._, 45:263 (1977).

11. Vogel, W. and Bauknecht, Th., Differential chromatid staining by in vivo treatment as a mutagenic test system, _Nature_, 260:448 (1976).

12. Marquardt, H. and Bayer, U., The induction of sister chromatid
     exchanges in the bone marrow of the Chinese hamster, I: The sen-
     sitivity of the system (methyl methane sulphonate), Mutation
     Res.56:169 (1977).

13. Allen, J.W., Schuler, C.F., Mendes, R.W. and Latt, S.A., A
     simplified technique for in vivo analysis of sister chromatid
     exchanges using 5-bromodeoxyuridine tablets, Cytogenet. Cell
     Genet. 18:231 (1977).

14. Carrano, A.V., Thompson, L.H., Lindl, P.A. and Minkler, J.L.,
     Sister chromatid exchange as an indicator of mutagenesis,
     Nature 271:551 (1978).

15. Perry, P. and Wolff, S., New Giemsa method for the differential
     staining of sister chromatids, Nature 251:156 (1974).

16. Lambert, B., Hansson, K., Lindsten, J., Sten, M. and Werelius,
     B., Bromodeoxyuridine-induced sister chromatid exchanges in
     human lymphocytes, Hereditas 83:163 (1976).

17. Lambert, B., Lindblad, A., Nordenskjöld, M. and Werelius, B.,
     Increased frequency of sister chromatid exchanges in cigarette
     smokers, Hereditas 88:147 (1978).

18. Crossen, P.E., Drets, M.E., Arrighi, F.E. and Johnston, D.A.,
     Analysis of the frequency and distribution of sister chromatid
     exchanges in human lymphocytes, Hum. Genetics 35:345 (1977).

19. Morgan, W.F. and Crossen, P.E., The incidence of sister chroma-
     tid exchanges in cultured human lymphocytes, Mutation Res.
     42:305 (1977).

20. Parrish, J.A., Fitzpatrick, T.B., Tannenbaum, L. and Pathak,
     M.A., Photochemotherapy of psoriasis with oral methoxsalen
     and longwave ultraviolet light, N. Engl. J. Med. 291:1207
     (1974).

21. Swanbeck, G., Thyresson-Hök, M., Bredberg, A. and Lambert, B.,
     Treatment of psoriasis with oral psoralens and longwave ultra-
     violet light: therapeutic results and cytogenetic hazards,
     Acta Dermatovener (Stockholm) 56:367 (1975).

22. Scott, B.R., Pathak, M.A. and Mohn, G.R., Molecular and genetic
     basis of furocoumarin reactions, Mutation Res. 39:29 (1976).

23. Lambert, B., Morad, M., Bredberg, A., Swanbeck, G. and
     Thyresson-Hök, M., Sister chromatid exchanges in lymphocytes
     from psoriasis patients treated with 8-methoxypsoralen and
     longwave ultraviolet light, Acta Dermatovener, (Stockholm),
     58:13 (1978).

24. Lambert, B., Ringborg, U., Harper, E. and Lindblad, A., In-
     creased frequency of sister chromatid exchanges in patients
     receiving chemotherapy against malignant disorders, Cancer
     Treatment Reports, 62: 1413 (1978).

25. Lambert, B., Ringborg, U. and Lindblad, A., Prolonged increase
     of sister chromatid exchanges in lymphocytes of melanoma pati-
     ents after CCNU treatment, Mutation Res., 59:295 (1979).

26. Funes-Cravioto, F., Zapata-Gayon, C., Kolmodin-Hedman, B.,
     Lambert, B., Lindsten, J., Norberg, E., Nordenskjöld, M.,
     Olin, R. and Swensson, Å., Chromosomal aberrations and sister
     chromatid exchange in workers in chemical laboratories and a
     rotoprinting factory and in children of women laboratory
     workers, Lancet II:322 (1977).

27. Funes-Cravioto, F., Zapata-Gayon, C., Kolmodin-Hedman, B.,
     Lambert, B., Lindsten, J., Norberg, E., Nordenskjöld, M.,
     Olin, R. and Swensson, Å., Chromosome aberrations and sister
     chromatid exchange in laboratory and factory workers and their
     children, in "Mutagen induced chromosome damage in man,"
     H.J. Evans and D.C. Lloyd (eds.) p. 275, Edinburgh University
     Press, Edinburgh (1978).

28. Ehrenberg, L., Hiesche, K.D., Osterman-Golkar, S. and
     Wennberg, J., Evaluation of genetic risks of alkylating agents:
     tissue doses in the mouse from air contaminated with ethylene
     oxide, Mutation Res., 24:83 (1974).

29. Calleman, C.J., Ehrenberg, L., Jansson, B., Osterman-Golkar, S.,
     Segerbäck, D., Svensson, K. and Wachtmeister, C.A., Monitoring
     and risk assessment by means of alkyl groups in haemoglobin in
     persons occupationally exposed to ethylene oxide, J. Toxicol.
     Environm. Health, (1978) (in press).

30. Lambert, B., Lindblad, A., Hansson, K., Calleman, C.J.,
     Ehrenberg, L. and Osterman-Golkar, S., Increased frequency of
     sister chromatid exchanges in human subjects exposed to ethylene
     oxide (1978) (to be published).

FAR AND NEAR ULTRAVIOLET LIGHT AS MOLECULAR PROBES FOR

ASSESSMENT OF DNA REPAIR AND RADIOSENSITIVITY

Emanuel Riklis

Israel Atomic Energy Commission, Radiobiology Dept.
Nuclear Research Center-Negev, Beer-Sheva, Israel

In previous lectures I have described the effects of radiation on cells and cell constituents, the methods of evaluating
the damage and the photo- and radiation- products which are
responsible for the damage, and lead to either cell death or,
what is sometimes worst, to mutation. For reasons of space and
scope there will be no attempt to repeat in detail the description of these effects, and the details may be read in the out-
standing presentation of B. Sutherland (1),and of Riklis (2),
as well as in the different papers in this volume dealing with
cellular repair mechanisms. If massive assaults by radiation
or other damaging agents may be recognized easily and measured
simply by survival curves, damage to the most sensitive cellular
constituent, DNA, requires biochemical methods of detection.
The lymphocyte, being the most sensitive cell, provided the
necessary object for studying effects, by cytogenetical methods,
but chromosome aberrations may be the end result of a series
of biochemical events, among them the most important is the
existence of molecular repair mechanisms, and it is the capacity
of repair which dictates the fate of an irradiated cell.
The phenomena of repair have become far more complicated than
when excision repair was first discovered in 1963 (3,4,5).
Not only repair capacity, but also repair fidelity determine
the ultimate fate of the cell and organism: excision repair
and photoreactivation are error-free, but other mechanisms
such as postreplication-recombination, and the recently discovered SOS repair are error-prone, thus increasing the chance
of mutation and carcinogenicity. In this respect, low-level
radiation becomes more disturbing as its effects are not
resulting so much in cell death as in cell mutation.

131

For many years research on low level radiation effects was
carried out with far ultraviolet light (FUV) and ionizing
radiations, and the link between photoproducts of DNA, or radiation products and chain breaks, and survival, was established.
Only recently, with the understanding that thymine dimers (TT)
are possible causes of skin cancer (6), attention turned to
focus on near ultraviolet (NUV) light and the effects of sunlight on cells. With the appearance of genetic auto-immune
diseases such as Xeroderma pigmentosum, or Ataxia telangectasia,
where susceptibility to UV light and gamma rays is apparent
respectively, the search of a link between these diseases
and DNA repair capability began, and soon turned out to be
of outmost importance, as such cells from patients can play
among mammalian cells the same role as played by mutant cells
in bacteria in which the basic studies of repair were done
so successfully. The term radiosensitivity should really be
changed now to radiation "susceptibility" and we understand
now that human cells show susceptibility to various agents
which affect DNA structure, and become "sensitive" to them
if and when the molecular repair mechanisms are impaired and
do not function normally. Thus people who carry the genetic
autoimmune disease Xeroderma pigmentosum (XP) are sensitive
to sunlight and are also cancer prone.

NUV LIGHT EFFECTS

The surge of interest in the possible harmful effects
of NUV or sunlight began only relatively recently along with
the marked increase in incidence of skin cancer and especially
melanomas. There is no direct evidence of the factor which
caused such an increase but it is commonly believed that modern
times habits of more leisure time, more travel to hot sunny
countries and increased exposure with less clothing and body
coverage are among the major reasons for the phenomena.
A gradual change in the spectrum of the sun's rays, bringing
about more energy at the wavelengths around 290 nm, is also
blamed as the cause for the increasing danger from the sun.
Sunburn is caused by UV rays under 320 nm, and so is skin
cancer, as the wavelengths which produce various photoproducts
are between 254 nm FUV to above 313 nm. At 365 nm however,
new types of products are formed and play the major role in
possible damages. These are products formed by sensitization
of DNA when molecules known as sensitizers are acting together
with NUV light, especially of 365 nm, to form crosslinks with
the DNA bases, monoadducts, or diadducts crosslinking the two
strands of a DNA molecule and impairing its biological activity.

The known effects of NUV light are the following:
Pyrimidine dimers are formed, but at a much smaller rate
than by 254 nm; the ratio of dimers formed by 254 to those
formed by 365 nm is $7x10^5$. This ratio is also similar to the
rate of inactivation of bacteria by these radiations.
Some single strand breaks are formed by 365 nm light ($3x10^{-5}$
breaks per genome per Jm^{-2}). There are certainly far more
breaks than dimers formed by 365 nm as compared to 254 nm (7).
The enzymatic process of photoreactivation is inhibited by
NUV light and dimer excision is also impaired by 10^6 J/m^2.
AET reduces breaks by a factor of 3 to 9, suggesting that
free radicals are involved in the production of breaks by NUV.
There are many other effects which do not involve damage to
DNA, such as membrane damage, transport and metabolic system
damage etc., but those are beyond the scope of this presentation.
What is of greater interest to our discussion are the effects
caused by NUV in combination with compounds which exert a
photosensitizing action. Of these furocoumarins, psoralen and
some of its derivatives, are particularly potent in forming
crosslinks in DNA, a fact which is utilized in a very interesting way for studying DNA repair.

APPLICATIONS OF PSORALEN PLUS NUV LIGHT

Since cells taken from patients with autoimmune diseases
showed impaired repair capabilities, the accurate assessment
of DNA repair is of utmost importance. The most popular and
commonly used method of assay is based on the measurement of
the rate of incorporation of labelled thymidine into DNA,
incorporation which indicates either semiconservative DNA
synthesis, or unscheduled repair synthesis. For repair, which
proceeds following exposure to radiation or other damaging
agents, to be measured, the normal replicative synthesis must
be stopped. This is achieved by several methods, the most
common is the addition of hydroxyurea as an inhibitor of semiconservative DNA synthesis. This method has drawbacks and
side effects, the most obvious of which is in the fact that
it still leaves several percent of the synthesis to proceed,
thus allowing for a background of several thousands of counts
of incorporated thymidine, masking the small number of counts
which are the result of true repair synthesis.

Exposure of Chinese hamster V79 cells or human skin
fibroblasts to TMP (4,5;8 trimethylpsoralen) plus NUV light
results in a drastic inhibition of replicative DNA synthesis
the extent of which depends on the exposure, TMP concentration,
and the length of time between the exposure to TMP plus NUV
and the addition of ^3H-thymidine. Optimum conditions bring

about inhibition of 99.85% of thymidine incorporation into
Chinese hamster cells and 99.5% into human fibroblasts.
The combination of 6000 Jm^{-2} NUV light exposure of cells with
$5x10^{-6}M$ TMP and a 3 hr period of interim incubation is there-
fore routinely employed by us (8), whenever studies require
cessation of replicative DNA synthesis.

Several significant results have already been obtained
with this new method: it enables very accurate measurement
of repair synthesis which takes place only after the cells
are exposed to an acute dose of radiation, or of MMS (methyl-
methanesulfonate) which mimicks gamma radiation.
Indeed, cells of normal human beings incorporated labelled
thymidine at a much higher rate than cells taken from Ataxia
patient, following exposure to either gamma radiation or MMS,
as can be seen in one of the examples shown in Fig. 1.
In this figure, B.B. is a normal healthy individual, while
PE is a boy who is under observation since his repair capacity
has been found impaired when tested by several methods. The
rate of incorporation of labelled thymidine as measured by
our method shows clearly the extent of this impairment.

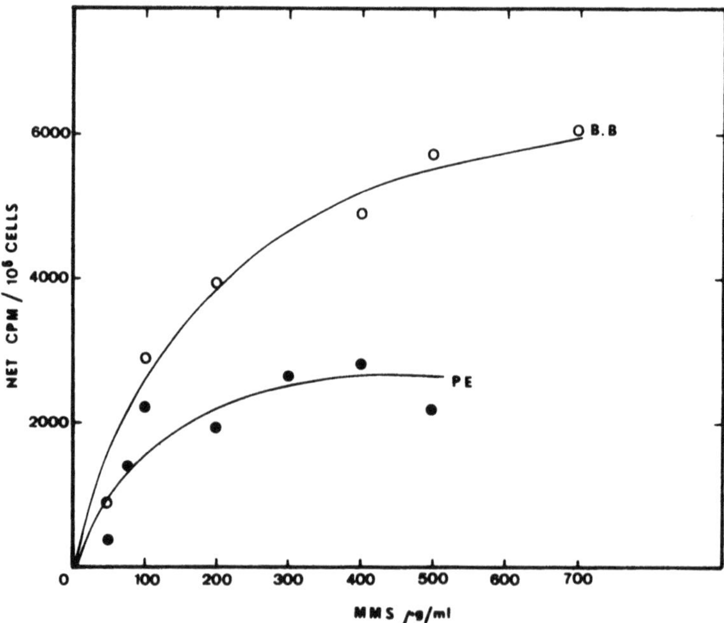

Fig. 1   Incorporation of H^{3}-TdR into fibroblasts
            treated with MMS.

In having a procedure which enables accurate assessment
of repair capacity, one may develop this to show not only
pre-existing impaired capacity, but also whether the normal
repair capacity is affected by chronic low-level radiations.
Indeed, such a study is being carried out in our laboratory (9)
with Chinese hamster cells which are being exposed to 20 Rads
per day gamma radiation. After three months of this treatment
there is no effect on survival of these cells as compared to
survival curves of control cells, but what is more significant
is that the rate of incorporation of thymidine through repair
synthesis is also functioning at normal capacity, although
the rate of normal DNA synthesis seems to be somewhat slower.
Continued development of this method may give us a much needed
method for accurate measurement not only of damage, but of
possible potential damage as expressed in impaired repair.

MUTAGENICITY

The fact that a cell survives an assault and continues
to perform DNA synthesis and even repair synthesis - still
is not sufficient proof of its health. It may be alive, yet
mutated by unrepaired damage which may lead to transformation
and to carcinogenicity. The method described of using TMP
plus NUV light may be useful also in developing a suitable
assay system which will provide an indication of the state
of health of the cell; this can be done if one measures not
simply incorporation of thymidine, but rather the activity
of certain induced enzymes. TMP plus NUV light treatment does
not interfere with the translation process and if transcription
is not damaged by the assault of acute radiation - translation
will proceed normally as can be seen, for example by studying
the induction of the enzyme ornithine decarboxylase in Chinese
hamster cells (10).

Another aspect of interest in the problem of assessment
of the effects of radiation and chemicals at low doses is
beginning to find its solution by the development of systems
which can test whether a compound is mutagenic, and as such
also potentially carcinogenic. The simplest and most widely
used assay is the Ames Test (11) based on reversion of certain
bacterial mutants from hystidine dependent to non-dependent.
It was of interest to study the extent of mutagenicity of
far UV and near UV radiations, alone and when employed in
combination with some of the sensitizing compounds mentioned
previously. Fig. 2 shows the results of such a study, and
it is clear from it that psoralen is a powerful mutagen,
when employed in combination with NUV light, while NUV alone

is only slightly mutagenic. Hematoporphyrin plus far UV is
not mutagenic whatsoever and chlorpromazine plus NUV, a drug
which causes photoallergic reactions - is somewhat mutagenic.
These results demonstrate the need to take the actions of
light radiation into consideration when testing the mutagenic
activities of various chemical compounds and signifies the
importance of these low energy radiations in our lives.

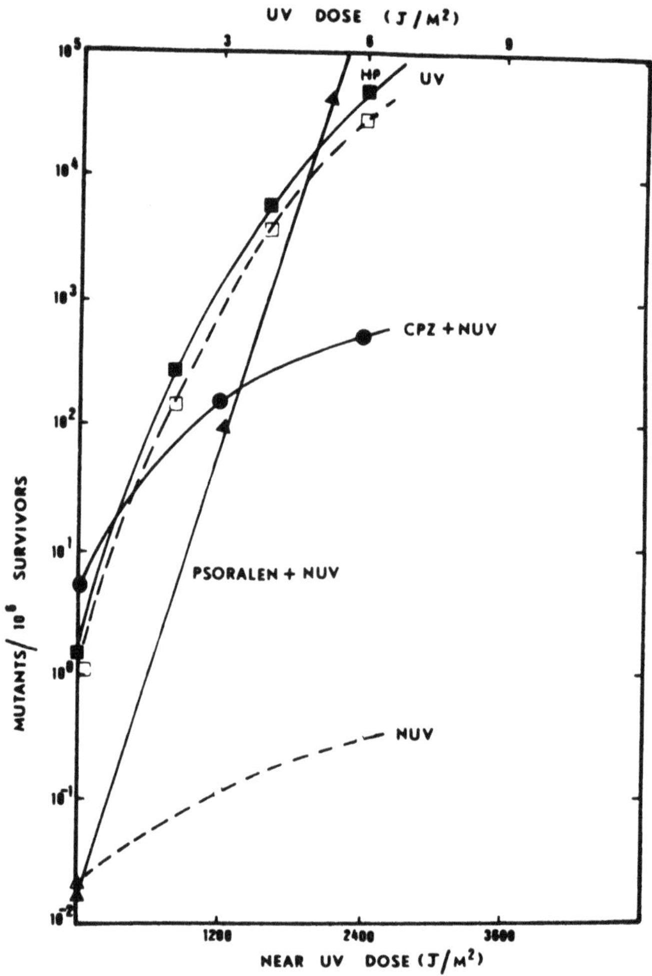

Fig. 2   The mutagenic activity of UV light in induc-
            ing histidine independence of S. typhimurium
            in presence and absence of photosensitizing
            drugs. The Ames method was used with some
            modification.
            Near UV, △ ; Near UV + TMP, ▲ ; Near UV +CPZ,
            ● ; far UV, ☐ ; far UV + HP, ▣ .

REFERENCES

1. Sutherland B.M., in "Synchrotron Radiation Applied to
   Biophysical, Biochemical and Biomedical Research"
   NATO Advanced Studies Institute Summer School,
   Plenum Publ. Corp. 1979.

2. Riklis E., ibid

3. Setlow R.B. and Carrier W.L., Proc. Nat. Acad. Sci.
   USA 51, 226 (1964).

4. Howard-Flanders P. and Boyce R.P., Proc. Nat. Acad.
   Sci. USA 51, 239 (1964).

5. Riklis E., Canad. J. Biochem. 43, 1207 (1965).

6. Setlow R.B., Proc. Nat. Acad. Sci. USA 71, 3363 (1974).

7. Elkind M.M., Han A. and Chang-Liu C.M., Photochem.
   Photobiol. 27, 709 (1978).

8. Heimer Y.M. and Riklis E., Biochemical J., in press.

9. Riklis E., Kol R. and Heimer Y.M., in preparation.

10. Heimer Y.M. and Riklis E., in "DNA Repair and Late Effects"
    IGEGM Intl. Meeting, Tel-Aviv 1978 (Editors: Riklis, Slor
    and Altmann).

11. Ames B.N., McCann J. and Yamasaki E., Mut. Res. 31,
    347 (1975).

ENZYMES FROM CALF THYMUS THAT MIGHT BE INVOLVED IN DNA REPAIR °

Francesco Campagnari

Laboratory of Biochemistry, Biology Group Ispra, D.G.XII,

C.E.C.,Joint Research Centre, 21020 Ispra (Va), Italy

1. DNA DAMAGE AND REPAIR

The DNA molecules carrying the genetic information of cells
are susceptible to damage by radiation of certain wavelengths,
such as X-, γ- and UV-rays (see for review: Latarjet, 1972; Blok
and Loman, 1974; Murphy, 1974), and by organic chemicals that themselves or through their metabolic products react with nucleic acids
as strong electrophiles (Miller and Miller, 1966; Miller, 1970).
Living cells exposed to these noxious agents undergo functional
inactivation, enhancement of the mutation rate and decrease of survival. In vertebrate animals the lesions of DNA may initiate complex processes of cell transformation that will eventually lead to
cancer. Radiation and substances that directly or indirectly alter
the molecules of DNA are potentially mutagenic and carcinogenic,
thus behaving as genetic toxins. Detrimental agents of this type
are natural or artificial components of the environment in which
we all presently live.

The DNA cannot be considered as an entirely stable compound.
Its primary structure is subjected with time to a number of hydrolytic degradations, even under physiological conditions of temperature and pH. The chemical decay of the heterocyclic base residues
in DNA becomes of practical importance for the large genomes of
resting and proliferating mammalian cells. Calculations from in

°This publication is contribution N° 1667 of the Programme Biology,
  Radiation Protection and Medical Research, Directorate General XII
  for Research, Science and Education of the Commission of the European Communities.

vitro measurements show that the depurination of dAMP and dGMP nucleotidyl units in the DNA of higher eukaryotes has a definite probability to occur (Lindahl and Nyberg, 1972). Very likely this is
equally true for the deaminations of 5-methylcytosine to thymine
and also of cytosine to uracil, at least in the dC:dG enriched sequences of the same DNAs (Lindahl and Nyberg, 1974; Lindahl, 1978).

During their lifetime, the cells strive to maintain the functional integrity of genetic material with the aid of serial enzymatic reactions that eliminate incidental lesions from the DNA molecules. These mechanisms of DNA repair are variously distributed in
all forms of life and are quantitatively modulated even among the
members of a same biological population. The actual capacity to repair DNA is the determinant of the individual sensitivity to the
various genetic injuries. Conceivably, DNA repair contributes also
to counteract the time dependent accumulation of DNA damages that
has been regarded as one of the causes of cellular ageing (Holliday
and Tarrant, 1972; Hart, 1976).

Most of our notions about the basic processes of DNA repair
come from work carried out in phage-infected or uninfected bacterial cells. Significant progresses were made after the discovery
by Ruth Hill of the first radiosensitive mutants in Escherichia
coli (Hill, 1958). Thereafter, many bacterial strains with mutations affecting the resistance to radiation and/or to genetic toxins
have been isolated. The proper use of crosses between proficient
and deficient cells and the genetic analysis of the progeny allowed
detection of distinct forms or even single reactions of DNA repair.
The informations gained from these genetic experiments greatly facilitated the identification of the competent enzymes. In fact six
years after the Hill's report, an integral process of DNA repair,
namely the removal of pyrimidine dimers from UV-irradiated DNA in
Escherichia coli, could be firstly described as a consistent sequence of enzymatic reactions (Setlow and Carrier, 1964). Up to
date other equivalent schemes for the repair of specific DNA injuries in bacteria have been proposed. We shall not elaborate on
these studies but we will use the derived conceptual models to
outline the general properties and the pattern of the repair
mechanisms that are supposed to operate in mammalian cells.

The processes of DNA repair are differentiated and extremely
selective in the initial steps brought about by enzymes that recognize only specific molecular damages or unique structures of repair intermediates in DNA. After the removal of altered DNA nulceo-
tides, the repair mechanisms may follow common routes and share biochemical reactions. Conversely, a given DNA lesion can be dealt with
by substitutive pathways and a same molecular modification of DNA
can be promoted by different enzymes.

It also appears that each system of DNA repair is under the
control of numerous genes whose polypeptide products are insuffi-
ciently known even for the most investigated Escherichia coli
(Clark and Ganesan, 1975) and are largely ignored for the highest
organisms (Grossman et al., 1975). This applies also to man, as
exemplified by the increasing number of mutated genes with unas-
signed functions that are being recognized in the repair deficient
hereditary disorder xeroderma pigmentosum (Bootsma, 1978; Keijzer
et al., 1979). It should be reminded that regulatory proteins for
transcriptional control of biochemical processes are rare, or perhaps undetectable, in mammalian cells and none of them with a
clear relation to DNA repair has been found as yet.

We may reasonably conclude that the mechanisms of DNA repair
in vivo are much more intricate and complex than their current description in biochemical terms. Especially for eukaryotic cells,
the approach to the biochemistry of DNA repair is mostly restricted
to define a minimal sequence of enzymatic reactions which is in
agreement with the experimental observations and can coherently explain the corrective changes for restoration of the DNA integrity.

2. DNA REPAIR IN MAMMALIAN CELLS

For a biochemical description, the processes of DNA repair in
mammalian cells can be ascribed to three general operations, namely
excision repair, post-replication repair and sealing of strand
breaks. The first two rely on multi-step mechanisms and the last
one is centered around a single enzymatic event. As a complement
to these basic pathways, auxiliary reactions promoted by phosphat-
ases, esterases and kinases can provide the altered DNA molecules
of the functional termini needed for the attack by repair enzymes.

In this presentation, we shall not consider the photoreactivation of thymine dimers in UV-irradiated DNA, a change mediated either by chemical sensitizers (Roth and Lamola, 1972) or by the
enzyme DNA photolyase that has been found also in placental mammals
(Sutherland et al., 1974).

2.1. Excision Repair

Excision repair is represented schematically in Figure 1 and
refers to a set of transformations in which damaged portions of DNA
are eliminated and substituted for by new stretches of the original
nucleotides. Through several variants, this mechanism acts on DNA
lesions of different origin and with distinct molecular features.
In mammalian cells, excision repair has been reported first for UV-
induced pyrimidine dimers of human DNA (Regan et al., 1968) and

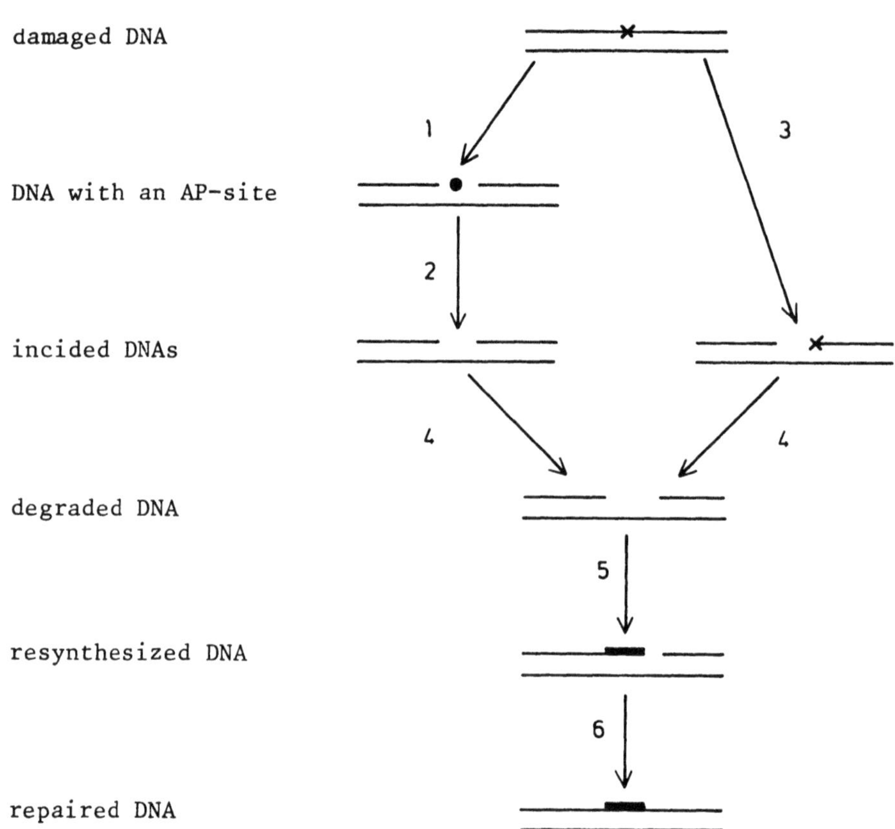

FIGURE 1. Minimal sequence of reactions for excision repair of damaged DNA. The numbered reactions are promoted by various types of enzymes as follows: 1. DNA glycosylases; 2. AP-endonucleases; 3. specific endonucleases; 4. exonucleases; 5. DNA polymerases; 6. DNA ligases.

subsequently for other DNA abnormalities, such as: the 5,6-dihydro-xy-dihydrothymine residues formed by ionizing radiation (Mattern et al., 1973); chemically methylated purines (Lawley, 1975); and base adducts produced by a number of carcinogens(Strauss, 1976). The altered nitrogenous bases can be removed from DNA either in the form of nucleotides or as free pyrimidines and purines. These two options have been termed as the nucleotide excision and base excision modes (Lindahl, 1976; see also Hanawalt, 1977).

The overall repair pathway consists of four concerted operations.

1) Incision of DNA near the damaged site. This is regarded as the committed step that is carried out by enzymes recognizing the given DNA lesions. It may occur either in a single endonucleolytic reaction by a specific endonuclease or in a two step process due to the combined action of a DNA glycosylase releasing the altered base and of an apurinic/apyrimidinic endonuclease which cleaves the sugar-phosphate backbone of DNA.

2) Degradation and excision of a nucleotide sequence by an exonuclease that forms a gap in the affected strand of DNA. The size of the gap varies according to the nature of the lesion and ranges from about one hundred nucleotides in UV-irradiated DNA (Cleaver, 1968; Regan et al., 1971) to apparently few base residues in DNA injuries by X-rays (Fox and Fox, 1973).

3) Repair replication by a DNA polymerase that fills the mono-helical gap with a new polydeoxynucleotide chain. DNA polymerization starts from the terminal nucleotidyl unit with a free 3'-OH group, at the edge of the gap, and proceeds to complement the unpaired portion of the opposite DNA strand, which is used as a template. After this step a simple monohelical nick with a $3'$-OH// $5'$-PO$_4$ conformation is left in the DNA molecule.

4) Sealing of the residual nick by DNA ligase. The enzyme rejoins the contiguous polydeoxynucleotide chains of the interrupted strand and restores DNA integrity.

Excision repair is basically non-mutagenic and error free because it allows accurate correction of structural alterations in DNA. In fact, the defective nucleotide sequence is degraded and resynthesized under the guidance of the intact copy printed in the antiparallel strand of DNA.

2.2. Post-replication Repair

Post-replication repair concerns DNA lesions that escape excision repair. The presence of molecular damages in replicating DNA
poses a temporary block to the machinery for DNA synthesis. Nascent
polydeoxynucleotide chains stop growing when physical chemical
changes are encountered in DNA templates. Gaps in the daughter DNA
strands form opposite these lesions and extend up to the next segments of replicated DNA. Afterwards however, DNA polymerization by-
passes the damage sites of the parental DNA templates and resumes
the assembling of the daughter strands up to completion.

In bacteria, post-replication repair develops with an exchange
of DNA material between parental and daughter strands (Rupp et al.,
1971) and this sort of illegitimate DNA recombination fills the gaps
in the newly synthesized polydeoxynucleotide chains. Recombinational
DNA exchanges seem to be infrequent or minimal in the post-replication repair of mammalian DNA (Lehmann, 1972) and alternative mechanisms have been considered (Lehmann, 1975); Higgins et al., 1976).
However, the post-replication repair of DNA in mammalian cells has
still not been described satisfactorily and the details of its biochemical pathway remain largely to be elucidated. It should be noted
that the bypass of DNA lesions during or after DNA replication in
mammalian cells share with excision repair both the filling of
single stranded gaps by polymerase and the closing of strand breaks
by ligase. (Figure 2).

The replication of DNA containing molecular alterations may
result in the formation of incorrect nucleotide sequences. Therefore, post-replication repair has been viewed mainly as error prone,
although some of the proposed mechanisms for such a repair imply
faithful duplication of the DNA strands (Higgins et al., 1976).

It has been suggested that a mutagenic post-replicative DNA
repair, called "SOS repair", can be induced in prokaryotes by
agents causing DNA lesions (Radman, 1974). The presence of a similar inducible error prone process of DNA repair in mammalian cells
has not been proved conclusively.

2.3. Sealing of Strand Breaks

Breaks in the sugar-phosphate backbone of DNA strands are produced with a high incidence by ionizing radiations (see: Latarjet,
1972; Blok and Loman, 1973), but they are also found after heavy
exposures to UV-light (Marmur et al., 1961) and treatment with
metal-binding antibiotics of the bleomycin type (Taheshita et al.,
1978). The nicks formed in DNA are not always hydrolytic scissions
of the phosphodiester bridges between the deoxynucleotides. Often,
they are associated with chemical changes of the deoxyribose resi-

Post-replication Repair

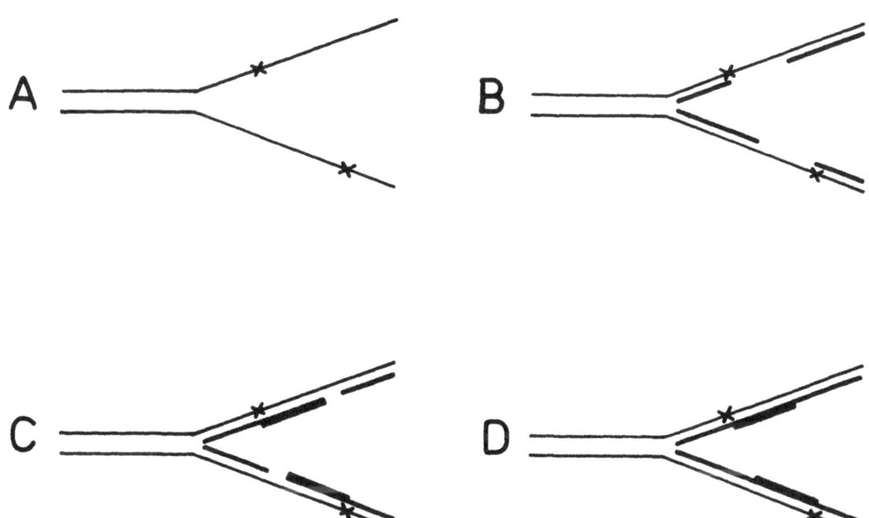

FIGURE 2.   Scheme of events in post-replication repair of
mammalian DNA.
      A. Opened replication fork with DNA damaged
      B. Replicated fork with gaps in daughter strands
      C. DNA polymerization bypassing lesions
      D. Ligation of newly formed DNA chains.

dues, release of the bases and even loss of terminal nucleotidyl
units. In the last instances, the sealing of strand breaks is
coupled with reactions of the excision repair pathway.

The enzyme DNA ligase is able to close directly the single
stranded breaks of DNA that have a $3'-OH//5'-PO_4$ conformation. For
other simple monohelical interruptions, the 3' and/or 5' chemical
termini in the nicks must be properly modified by auxiliary enzymatic reaction before ligation takes place.

The rejoining enzyme shows an absolute specificity for justa-
posed $3'-OH$ and $5'-PO_4$ termini in the sealable nicks, but it has a
loose requirement for the integrity of the nucleotide sequences in
the broken DNA chains. Sealing of monohelical scissions in DNA is
a non-limiting process for mammalian cells and occurs also after
supra-lethal doses of ionizing rays (Lett et al., 1967). Surely,
much of the rejoining in the irradiated cells must have been carried out on damaged DNA and has to be ascribed to misrepair.

The ligase property to act also on DNA molecules with structural alterations is of practical importance. In fact, ligation of
an interrupted DNA strand still endowed with nucleotide lesions
might well facilitate the fixation of premutational changes in the
genetic material. Since the closing of a DNA nick is the last step
of functional DNA repair, mechanisms for elimination of base damage
from DNA in a concerted manner with the sealing of strand breaks
should operate in vivo. However, the control of DNA repair as to
prevent untimely ligations has not been studied in animal organisms.

The reported rejoining of double stranded scissions in mammalian DNA (Corry and Cole, 1973) may reflect the relaxed requirement of ligase for the spatial structure of DNA substrates. Occasional blunt joining of contiguous bihelical DNA fragments will
appear as a repair of double stranded breaks and this may well
occur for the folded DNA molecules of nuclear chromatin. However,
the relevance of such processes for reconstitution of a functional
genome and/or induction of chromosomal aberrations remains to be
established.

3. ENZYMES OF DNA REPAIR FROM CALF THYMUS

Bovine thymus represents a sort of model tissue for the study
of mammalian enzymes that act on DNA substrates and DNA templates.
In fact, it is the single mammalian source from which the largest
number of these functional proteins have been prepared and characterized. A short survey of the calf thymus enzymes that may have a
role in DNA repair pathways, here conceived as minimal sequences
of possible biochemical reactions, will provide a useful framework
of knowledge.

The four main classes of enzymes that have been implicated in
the repair of mammalian DNA, namely DNA glycosylases, DNases, DNA
polymerases and DNA ligases, are all present in calf thymus and
enzymes of each class have been the target of detailed investigations.

3.1. DNA Glycosylases

These glycosylases remove directly damaged or incorrect bases
from DNA and are the first components of base excision systems for
DNA repair. Each enzyme has a narrow specificity for a given ab-
normal base and cleaves the sugar-base glycosyl linkage of an im-
proper deoxynucleotidyl unit in DNA, thus releasing the base and
leaving an apurinic-apyrimidinic site. This site is recognized by
a specific DNase, AP-endonuclease, which hydrolyzes its phosphodiester bond and produces a single strand break with partial or
total loss of a nucleotidyl residue. After the strand scission, the
structural integrity of DNA might be recovered through the series
of exonuclease, DNA polymerase and DNA ligase reactions occurring
after the initial incision step in the excision repair process.

The uracil-DNA glycosylase from calf thymus has been studied
in our laboratory. The enzyme was partially purified from extracts
of thymocyte nuclei and characterized. It catalyzes the release of
uracil from synthetic DNAs containing uracil residues, uracil-DNAs,
and has an estimated molecular weight of 28,700. The glycosylase
reaction does not require cofactors nor divalent metal cations,
occur on polymeric substrates and not on dUMP, and is clearly of
the hydrolytic type with an equilibrium greatly shifted towards
the dissociation of uracil from DNA (Talpaert-Borlé et al., 1979).
With the exception of the preference for double stranded uracil-
DNAs as substrates and of the higher molecular weight, the enzyme
properties are similar to those reported for the analogous bacterial enzymes by other investigators (Cone et al., 1977; Lindhal
et al., 1977).

The obvious role of this glycosylase is to remove from DNA the
uracil which is formed by deamination of cytosine or in incorporated by polymerization of dUTP instead of dTTP substrates. The
enzyme is rather ubiquitous and has been obtained from a number of
animal sources. We have detected uracil-DNA glycosylase activity
in extracts from the main parenchimatous tissues of rat, the highest
concentration being found in thymus (Talpaet-Borlé and Campagnari,
unpublished results).

Other DNA glycosylases for base excision have not been carefully investigated in mammalian cells. Although some of the reported
reactions might need better identification, circumstantial evidence

for glycosylases active on alkyl-DNAs has been forwarded (Lawley, 1975). An enzyme releasing 3-methyladenine from alkylated DNA has been recovered from cultured human lymphoblasts (Brent, 1979). This glycosylase does not require divalent metal cations and has an estimated molecular weight of 34,000. Therefore, we may consider that at least one alkyl-DNA glycosylase, namely the enzyme acting on 3-methyladenine-DNA, is functionally expressed in mammalian lymphoid tissues, including thymus.

3.2. AP-endonuclease

An enzyme which introduces single strand breaks at the apurinic-apyrimidine sites of bihelical DNA has been isolated from calf thymus. It has a molecular weight in the 30,000-35,000 range and displays optimal catalysis at pH 8.5 and in the presence of Mg^{++} or Mn^{++} (Ljungquist et al., 1974a). The AP-endonuclease cleaves DNA at the 3' side of the bare deoxyribose residue with the release of a $5'-PO_4$ nucleotide terminus. It is active on heat depurinated DNA and on irradiated DNA with alkali labile sites that correspond to base losses in single nucleotidyl units (Ljungquist et al., 1974b).

This enzyme is a monofunctional protein without associated phosphatase and exonuclease activities and can be considered as representative of mammalian AP-endonucleases. It is at variance with the main AP-endonuclease of bacterial cells, namely endonuclease II, which performs multiple catalytic functions and should be better termed as exonuclease III* (Weiss, 1976).

3.3. Endonucleases

Early reports indicate that a DNase activity of type II is present in nuclei of calf thymus cells (Slor and Lev, 1971). This enzyme would be similar to the lysosomal acid DNase which degrades DNA to oligonucleotides by promoting hydrolytic chain scissions with a $3'-PO_4//5'-OH$ conformation (Bernardi, 1971). The nuclear form of calf thymus DNase II has not been clearly identified and its function, especially with regard to DNA repair, is rather dubious.

Another endonuclease, that is also detectable in cell nuclei but might well be involved in DNA repair, has been purified to homogeneity from bovine thymus and has been distinctly characterized (Wang and Furth, 1977). It is a tetrameric protein composed of identical subunits with molecular weights of 13,000 and possessing an exceptionally high isoelectric point. The enzyme shows a pH optimum at 6.6 and a requirement for Mg^{++} or Mn^{++}; it hydrolyzes

extensively single stranded DNA but introduces only a limited
number of monohelical nicks into native DNA. It is unknown whether
this limited attack on DNA reflects a specificity for rare base
sequences or for casual changes in the structure of double stranded
DNA. Apparently, this endonuclease has several properties that are
expected attributes of a DNA inciding enzyme for repair. Such an
enzyme has been named DNase V and differs from the pancreatic
DNase I, which also forms $3'-PO_4$ strand breaks but degrades native
DNA almost to completion (Laskowsky, 1971).

An example of mammalian endonuclease specific for DNA lesions
is provided by another calf thymus DNase, which has been partially
purified and found to hydrolyze internal phosphodiester bonds in
irradiated double stranded DNAs (Bacchetti and Benne, 1975). This
enzyme is active also in the absence of divalent cations, recognizes
apparently unidentified chemical alteration caused in DNA by either
UV-light or γ-rays, and releases $5'-PO_4$ termini at the incision
sites.

Many of the endonucleases for DNA repair that have been noted
in various mammalian tissues can be safely assigned to the DNase
types isolated from calf thymus.

3.4. Exonucleases

After the enzymatic incision of damaged DNA, the defective
nucleotidyl residues in or nearby the strand break should be removed
by exonucleases. These enzymes have not been investigated in calf
thymus glands as the endonucleases. However, two exonucleases,
namely DNase III and DNase IV, have been isolated from rabbit bone
marrow by Lindhal and coworkers and will be shortly described. Both
are Mg^{++}-dependent nucleases localized in cell nuclei.

DNase III has an estimated molecular weight of 52,000 and
hydrolyzes sequentially polydeoxynucleotide chains from the 3' ends
thus liberating 5'-mononucleotides and dinucleotides. Although it
has a preference for single stranded DNA as substrate, it degrades
at appreciable rates also native DNA (Lindhal et al., 1969a).

DNase IV has an approximate molecular weight of 42,000 and
promotes exonucleolytic degradation of double stranded DNA from
the 5' ends with release of 5-mononucleotides. The enzyme appears
to be specific for bihelical DNA substrates and hydrolyzes the first
nucleotides in DNA at higher rates than the subsequent ones
(Lindhal et al., 1969b).

Exonucleases similar to DNase III and DNase IV are widely
distributed and can be detected as contaminants of other enzyme

preparations from nuclei of mammalian cells. As a matter of fact,
we reported the presence of DNase III activity in side fractions
obtained during the purification of a DNA polymerase from the nuclei
of calf thymocytes (Bekkering-Kuylaars and Campagnari, 1974).

It should be mentioned that a 3' → 5' exonuclease with "proofreading" function for DNA synthesis has been found associated with
DNA polymerase α in partially purified preparations from nuclei of
calf thymus cells. This 3' → 5' exonuclease excises in vitro mismatched nucleotides from the 3'-OH termini of synthetic DNA primers
in a coordinate manner with the DNA polymerase reaction; such an
activity is lost in the advanced stages of purification when the
polymerase seems to separate into low molecular weight polypeptides
(Clerici et al., 1980).

It is feasible that exonucleases present in cell nulcei and
active on double stranded DNA, such as DNase III, DNase IV and the
"proofreading" enzyme, exert repair functions on th genetic
material in specific occurrences.

3.5. DNA polymerases

The two main DNA polymerases found in calf thymus glands,
namely polymerase α and polymerase β, have been largely characterized by Bollum and coworkers and the gained knowledge has inspired
most of the subsequent work on the DNA synthesizing enzymes from
mammalian cells (see: Bollum F.J., 1975).

The DNA polymerase α of bovine thymus has been originally
purified and is usually recovered from cytoplasmic fractions
(Yoneda and Bollum, 1965; Bollum et al., 1974), but it is present
in appreciable quantities also in cell nuclei from which has been
extracted by mild procedures and purified (Bekkering-Kuylaars
and Campagnari, 1972). The enzyme is heterogeneous and has been
obtained in a number of forms with molecular weights ranging from
over 230,000 to about 130,000 (Momparler et al., 1973; Bollum et
al., 1974; Holmes et al., 1976). This reflects the breaking down
of either an enzyme aggregate or a functional complex during isolation. Extensive purification of both cytoplasmic and nuclear enzymes
gives rise to unstable or inactive forms with molecular weights
well below 100,000 (Bollum et al., 1974; Yoshida et al., 1974;
Bekkering-Kuylaars and Campagnari, 1974; Holmes et al., 1979). A
polypeptide or enzyme fragment with molecular weight of 70,000 has
been indicated as the active subunit common to all species of DNA
polymerase α (Bollum et al., 1974; Bollum, 1975).

In the presence of dNTP substrates, the polymerase α promotes
DNA synthesis on various primer template systems, such as self-

primed single stranded DNA, native DNA with monohelical gaps and
synthetic DNA templates paired with either deoxyribonucleotide or
ribonucleotide primers. Its ability to utilize RNA chains as primers
has been related to a role in initiating DNA replication in vivo
(Chang and Bollum, 1972). In fact, this enzymatic activity is always
high in tissues with many dividing cells, like thymus, and has been
found to increase in response to stimuli for cell proliferation
(Bollum, 1975).

The α-polymerase shows pH optimum for activity at neutrality,
requires Mg^{++} or Mn^{++} as metal cofactors respectively with DNA or
homopolydeoxynucleotide templates, and is inhibited by concentrations of 0.1 M salts and of 1 mM N-ethylmaleimide. The enzyme catalysis on the primer-templates occurs in a distributive manner through
short rounds of nucleotide polymerization and not by formation of
a long DNA chain at a single time (Chang, 1975). Only a trace level
of 3' → 5' "proofreading" exonuclease activity has been detected
in the calf thymus DNA polymerase isolated from cytoplasmic fractions, whereas an error correcting exonuclease seems to be present
in partially purified preparations of the DNA polymerase α extracted
from cell nuclei as already mentioned (Clerici et al., 1980).

With regard to the enzyme activity on damaged DNA, it appears
that the complementary duplication of the template chain is induced
at normal rates on mildly X-irradiated DNA (Campagnari et al.,
1967) but is blocked and does not go to completion on DNAs exposed
to heavy doses of UV-light (Bollum and Setlow, 1963) or to X-ray
doses over 2,000 and up to 10,000 rads (Campagnari and Bertazzoni,
1968).

The DNA polymerase β of calf thymus is mainly bound to the
nuclear chromatin, from which it has been purified to homogeneity,
and is detectable also in cytoplasmic fractions. The enzyme consists of a basic polypeptide with a molecular weight of 45,000 and
acts in the presence of suitable templates, complementary dNTPs and
Mg^{++} or Mn^{++} as divalent metals. Its catalytic properties are similar with those of the α-polymerase, but it has a rather alkaline
pH optimum, is insensitive to N-ethylmaleimide and is stimulated,
not inhibited by 0.1-0.2 M salts (Chang, 1973a). Moreover at variance with the enzyme α, this deoxynucleotidyl-transferase induces
DNA synthesis only by elongating DNA primers (Chang and Bollum,
1972), while being able to use effectively RNA chains as templates
(Bekkering-Kuylaars and Campagnari, 1974; Bollum, 1975). Although
the purified DNA polymerase β does not have an associated 3' → 5'
exonuclease with error correcting function, it has been found to
replicate complementarily the templates with high accuracy and with
a mistake frequency below 10^{-4} (Chang, 1973b).

The amount of this enzyme in bovine thymus is low and falls

within the narrow range of values observed in other mammalian organs in terms of units/g of tissue. The DNA polymerase is present in all cells at almost the same constant concentration, quite independently upon metabolic changes and proliferative states (Bollum, 1975). The properties and the distribution of this enzyme are suggestive evidence for its role in DNA repair. However, it is unknown whether DNA repair synthesis can be alternatively or substitutively carried out also by the polymerase α in the late G_1, S and early G_2 phases of the cell cycle, when high levels of this DNA replicative enzyme are available.

3.6. DNA ligase

Like other eukaryotic ligases, the DNA-joining enzyme partially purified from calf thymus uses ATP-Mg as an energy donor to seal single strand breaks in native DNA and has an absolute requirement for the presence of adjacent 3'-OH and 5'-PO_4 termini at the nicks (Bertazzoni et al., 1972).

As already anticipated, mammalian ligases are rather unspecific for the correct structure of DNA outside the internucleotide scission. In fact, the calf thymus enzyme has been found capable to rejoin breaks in natural and synthetic DNAs, previously exposed to X-rays and thus containing significant radiation damage (Mathelet et al., 1974). In these experiments, all the nicks with the expected 3'-OH//5'-PO_4 conformation were sealed by ligase, although the alterations in the nucleotides close to the breaks prevented some specific marker enzymes from recognizing the functional 3'-OH termini of the broken DNA chains. It is well possible that misrepair by ligation is more frequent in mammalian cells than in bacteria, which possess a NAD^+-dependent ligase with more stringent catalytic requirements (Lehman, 1974).

The above observations have been made with a ligase purified about 3000-fold over the thymus homogenate and corresponding to the main enzyme form which has been indicated as DNA Ligase I. According to data from various mammalian tissue, this is a large enzyme composed of various subunits and increasing in cells during DNA replication (Soderhall and Lindhal, 1976).

The significance of the minor enzyme form called DNA ligase II, which is more labile and has a lower molecular weight than DNA ligase I, is still doubtful. This enzyme has been extracted also from calf thymus, where it represents less than 7% of the total ligase activity, and has been found not to react with antisera prepared against purified DNA ligage I (Söderhäll and Lindhal, 1975).

4. CONCLUDING REMARKS

An inventory of the enzymes isolated from calf thymus and shown to act on different DNA structures has been performed with reference to the general schemes of DNA repair in mammalian cells. The survey has been forcibly restricted to the few ascertained reactions occurring in vitro on DNA molecules dispersed in aqueous solutions. Therefore, the direct attack of enzymes on DNA packaged with histones and nuclear proteins in the chromatin has not been considered, although this functional interaction determines the success or the failure of the DNA repair events in the cell nuclei (Smerdon et al., 1978). Moreover, the absence of specific informations for the thymus tissue has given rise to other omissions. In fact, the repair of mitochondrial DNA and the newly discovered enzymatic activities, such as that of the putative purine-DNA insertase isolated from human cells (Deutsch and Linn, 1979), have not been taken into account.

Nevertheless, the exercise of confronting established enzyme properties with potential functions in DNA repair might have been useful. The following conclusions can be warranted.

a) For mammalian DNA repair, a wide gap exists between the scarce notions on the enzymes involved and the large body of evidence for the numerous and complex pathways that operate in cells (see for review: Cleaver, 1978). A comprehensive biochemical description of DNA repair in mammalian cells is still premature, the only reasonable approach being the attempt to define minimal sequences of repair reactions. At the present time, the main contributions in this area by biochemists may well come from the work directed to characterize the action of single enzymes on DNA integrated into chromatin and to identify new types of enzymatic activities.

b) The enzymes of DNA repair perform vital functions and are indispensable. Their concentration in mammalian cells fluctuate within narrow ranges of values with the exception of the increases noted during the proliferative state for those activities that are involved also in DNA replication. It is unlikely that mammalian zygote, not endowed with a full complement of DNA repair reactions, can develop through the stages of embryonic and fetal life and will give rise to a viable organism.

Therefore, we may expect to find rarely humans that are born with truly enzymatic defects of DNA repair. This is much at variance with the frequent enzymatic etiology of many inborn errors of metabolism. Generally, in fact, the relevant enzyme activities are at almost normal levels in the cells from the inherited human diseases

with defective process of DNA repair. So far the only genetic deficiency of a repair enzyme might have been detected in a patient
with xeroderma pigmentosum of complementation group D, whose cells
had one-sixth of the activity normally displayed by an Ap-endonuclease (Kuhnlein et al., 1976).

c) The enzymes acting on DNA are essential component of the
mechanisms for DNA repair but are not the limiting factors in the
functional operating of such mechanisms inside mammalian cells.
Non-enzymatic gene products controlling or facilitating the biochemical reactions on nuclear chromatin are probably responsible
for most of the inborn defects of DNA repair in man.

REFERENCES

Bacchetti, S., and Benne, R. (1975) Purification and characterization of an endonuclease from calf thymus acting on irradiated DNA,
Biochem. Biophys. Acta 390, 285-297.

Bekkering-Kuylaars, S.A.M., and Campagnari, F. (1974) Characterization and properties of a DNA polymerase partially purified from the
nuclei of calf thymus cells, Biochem. Biophys. Acta 349, 277-295.

Bekkering-Kuylaars, S.A.M., and Campagnari, F. (1972) Purification
of a DNA polymerase from calf thymus nuclei, Biochem. Biophys.
Acta 272, 526-538.

Bernardi, G. (1971) Spleen acid deoxyribonuclease, in "The Enzymes"
(P.D. Boyer, ed.), vol.4, pp.271-287, Academic Press, New York.

Bertazzoni, U., Mathelet, M., and Campagnari, F. (1972) Purification
and properties of a polynucleotide ligase from calf thymus glands,
Biochem. Biophys. Acta 287, 404-414.

Blok, J., and Loman, H. (1973) The effects of γ-radiation in DNA,
Current Topics Radiat. Res. Quart. 9, 165-245.

Bollum, F.J., Chang, L.M.S., Tsiapalis, C.M., and Dorson, J.W.
(1974) Nucleotide polymerizing enzymes from calf thymus glands,
Methods Enzymol. 29, 70-81.

Bollum, F.J. (1975) Mammalian DNA polymerases, Progr. Nucl. Acid
Res. Mol. Biol. 15, 109-114.

Bollum, F.J., and Setlow, R.B. (1963) Ultraviolet inactivation of
DNA primer activity. I. Effects of different wavelengths and doses,
Biochem. Biophys. Acta 68, 599-607.

Bootsma, D. (1978) Xeroderma pigmentosum, in "DNA Repair Mechanisms"
(P.C. Hanawalt, E.C. Friedberg, and C.F. Fox, eds.) pp. 589-601,
Academic Press, New York.

Brent, T.P. (1979) Partial purification and characterization of a human 3-methyladenine-DNA glycosylase, Biochemistry 18, 911-916.

Campagnari, F., Bertazzoni, U., and Clerici, L. (1967) The priming activity of X-irradiated deoxyribonucleic acid for the deoxyribonucleic acid polymerase from calf thymus, J. Biol. Chem. 242, 2168-2171.

Campagnari, F., and Bertazzoni, U. (1968) Effect of low doses of X-rays on isolated deoxyribonucleic acid (DNA). II. Results obtained by enzymatic methods, Medic. Nucl. Radiobiol. Latina 3, Suppl. 3, 304-311.

Chang, L.M.S., and Bollum, F.J. (1973) A comparison of associated enzyme activities in various deoxyribonucleic acid polymerases, J. Biol. Chem. 248, 3398-3404.

Chang, L.M.S., and Bollum, F.J. (1972) A chemical model for transcriptional initiation of DNA replication, Biochem. Biophys. Res. Commun. 46, 1354-1360.

Chang, L.M.S. (1973) Low molecular weight deoxyribonucleic acid polymerase from calf thymus chromatin. I. Preparation of homogeneous enzyme, J. Biol. Chem. 248, 3789-3795.

Chang, L.M.S. (1973) Low molecular weight deoxyribonucleic acid polymerase from calf thymus chromatin. II. Initiation and fidelity of homopolymer replication, J. Biol. Chem. 248, 6983-6992.

Chang, L.M.S. (1975) The distributive nature of enzymatic DNA synthesis, J. Mol. Biol. 93, 219-235.

Clark, A.J., and Ganesan, A. (1975) List of genes affecting DNA metabolism in Escherichia coli, in "Molecular Mechanisms of Repair of DNA" (P.C. Hanawalt, and R.B. Setlow, eds.) pp. 431-437, Plenum Press, New York.

Cleaver, J.E. (1968) Defective repair replication of DNA in xeroderma pigmentosum, Nature 218, 652-656.

Cleaver, J.E. (1978) DNA repair and its coupling to DNA replication in eukaryotic cells, Biochem. Biophys. Acta 516, 489-516.

Clerici, L., Sponza, G., Talpaert-Borlé, M., and Campagnari, F. (1980) Proofreading exonuclease activity in crude and partially purified preparations of DNA polymerase α from calf thymus, in "DNA Repair and Late Effects" (E. Riklis, H. Altman, and H. Slor, eds.) pp. 23-30, Nuclear Research Center Negev Publisher, Israel.

Cone, R., Duncan, J., Hamilton, L., and Friedberg, E.C. (1977) Partial purification and characterization of a uracil DNA N-glycosydase from Bacillus subtilis, Biochemistry 16, 3194-3201.

Corry, P.M., and Cole, A. (1973) Double strand rejoining in mammalian DNA, Nature New Biol. 245, 100-101.

Deutsch, W.A., and Linn, S. (1979) DNA binding activity from
cultured human fibroblasts that is specific for partially depurinated DNA and that inserts purines into apurinic sites, Proc.
Nat. Acad. Sci. USA 76, 141-144.

Fox, M., and Fox, B.W. (1973) Repair replication in X-irradiated
lymphoma cells in vitro, Int. J. Radiat. Biol. 23, 333-358.

Grossman, L., Braun, A., Feldberg, R., and Mahler, I. (1975)
Enzymatic repair of DNA, Ann. Rev. Biochem. 44, 19-43.

Hanawalt, P.C. (1977) DNA repair processes: an overview, in "DNA
Repair Processes" (W.W. Nichols, and D.G. Murphy, eds.) pp. 1-19,
Symposia Specialists, Inc., Miami.

Hart, R.W. (1976) Role of DNA repair in aging, in "Aging, Carcinogenesis and Radiation Biology" (K.C. Smith, ed.) pp. 537-556,
Plenum Press, New York.

Higgins, N.P., Kato, K., and Strauss, B. (1976) A model for replication repair in mammalian cells, J. Mol. Biol. 101, 417-425.

Hill, R.T. (1958) A radiation-sensitive mutant of Escherichia coli,
Biochem. Biophys. Acta 30, 636-637.

Holliday, R., and Tarrant, G.M. (1972) Altered enzyme in ageing
human fibroblasts, Nature 238, 26-30.

Holmes, A.M., Hesslewood, I.P., and Johnston, I.R. (1976) Evidence
that DNA polymerase of calf thymus contains a subunit of molecular
weight 155,000, Eur. J. Biochem. 62, 229-235.

Keijzer, W., Jaspers, N.G.J., Abrahams, P.J., Taylor, A.M.R.,
Arlett, C.F., Zelle, B., Takebe, H., Kinmont, P.D.S., and Bootsma,
D. (1979) A seventh complementation group in excision-deficient
xeroderma pigmentosum, Mutation Res. 62, 183-190.

Kuhnlein, U., Penhoet, E.E., and Linn, S. (1976) An altered apurinic
DNA endonuclease activity in group A and group D xeroderma pigmentosum fibroblasts, Proc. Nat. Acad. Sci. USA 73, 1169-1173.

Laskowski, M. Sr. (1971) Deoxyribonuclease I, in "The Enzymes"
(P.D. Boyer, ed.), vol. 4, pp. 289-311, Academic Press, New York.

Latarjet, R. (1972) Interaction of radiation energy with nucleic
acids, Current Topics Radiat. Res. Quart. 8, 1-38.

Lawley, P.D. (1975) Excision of bases from DNA methylated by carcinogens in vivo and its possible significance in mutagenesis and
carcinogenesis, in "Molecular Mechanisms for Repair of DNA" (P.C.
Hanawalt, and R.B. Setlow, eds.) pp. 25-28, Plenum Press, New York.

Lehman, I.R. (1974) DNA ligase: structure, mechanism, and function,
Science 186, 790-797.

Lehmann, A.R. (1975) Postreplication repair of DNA in UV-irradiated
mammalian cells, in "Molecular Mechanisms for Repair of DNA" (P.C.

Hanawalt, and R.B. Setlow, eds.) pp. 617–623, Plenum Press, New York.

Lehmann, A.R. (1972) Postreplication repair of DNA in ultraviolet-irradiated mammalian cells, J. Mol. Biol. 66, 319–337.

Lett, J.T., Caldwell, I., Dean, C.J., and Alexander, P. (1967) Rejoining of X-ray induced breaks in the DNA of leukaemia cells, Nature 214, 790–792.

Lindahl, T., and Nyberg, B. (1972) Rate of depurination of native deoxyribonucleic acid, Biochemistry 11, 3610–3618.

Lindahl, T., and Nyberg, B. (1974) Heat-induced deamination of cytosine residues in deoxyribonucleic acid, Biochemistry 13, 3405–3410.

Lindahl, T. (1979) DNA glycosylases, endonucleases for apurinic/apyrimidinic sites and base excision-repair, Progr. Nucl. Acid Res. Mol. Biol. 22, 135–192.

Lindahl, T. (1976) New class of enzymes acting on damaged DNA, Nature 259, 64–66.

Lindahl, T., Ljungquist, S., Siegert, W., Nyberg, B., and Sperens, B. (1977) DNA N-glycosydases. Properties of uracil-DNA glycosydase from Escherichia coli, J. Biol. Chem. 252, 3286–3294.

Lindahl, T., Gally, J.A., and Edelman, G.M. (1969a) Properties of deoxyribonuclease III from mammalian tissues, J. Biol. Chem. 244, 5014–5019.

Lindahl, T., Gally, J.A., and Edelman, G.M. (1969b) Deoxyribonuclease IV: a new endonuclease from mammalian tissues, Proc. Nat. Acad. Sci. USA 62, 597–603.

Ljungquist, S., and Lindahl, T. (1974a) A mammalian endonuclease specific for apurinic sites in double-stranded deoxyribonucleic acid. I. Purification and general properties, J. Biol. Chem. 249, 1530–1535.

Ljungquist, S., Andersson, A., and Lindahl, T. (1974b) A mammalian endonuclease specific for apurinic sites in double-stranded deoxyribonucleic acid. II. Further studies on the substrate specificity, J. Biol. Chem. 249, 1536–1540.

Marmur, J., Anderson W.A., Matthews, L., Berns, K., Gajewska, E., Lane, D., and Doty, P. (1961) The effects of ultraviolet light on the biological and physical chemical properties of deoxyribonucleic acids, J. Cell Comp. Phys. 58, Suppl. 1, 33–55.

Mathelet, M., Clerici, L., Campagnari, F., and Talpaert-Borlé, M. (1978) The activity of mammalian polynucleotide ligase on X-irradiated DNAs, Biochem. Biophys. Acta 518, 138–149.

Mattern, M., Hariharan, P.V., Dunlap, B.E., and Cerutti, P.A.
(1973) DNA degradation and excision repair in γ-irradiated Chinese
hamster ovary cells, Nature New Biol. 245, 230-232.

McKune, K., and Holmes, A.M. (1979) Further studies on partially
purified calf thymus DNA polymerase, Nucl. Acids Res. 6, 3341-3352.

Miller, E.C., and Miller, J.A. (1966) Mechanisms of chemical carcinogens: nature of proximate carcinogens and interactions with
macromolecules, Pharmac. Rev. 18, 806-838.

Miller, J.A. (1970) Carcinogenesis by chemicals: an overview, Cancer
Res. 30, 559-576.

Momparler, R.L., Rossi, M., and Labitan, A. (1973) Partial purification and properties of two forms of deoxyribonucleic acid polymerase from calf thymus, J. Biol. Chem. 248, 285-293.

Murphy, T.M. (1974) Nucleic acids: interaction with solar U.V.
radiation, Current Topics Radiat. Res. Quart. 10, 199-228.

Radman, M. (1974) Phenomenology of an inducible mutagenic DNA repair
pathway in Escherichia coli: SOS repair hypothesis, in "Molecular
and Environmental Aspects of Mutagenesis" (L. Prakash, F. Sherman,
M.W. Miller, C.W. Lawrence, and H.W. Taber, eds.) pp. 128-140,
C. Thomas Publ., Inc., Springfield.

Regan, J.D., Trosko, J.E., and Carrier, W.L. (1968) Evidence for
excision of ultraviolet-induced pyrimidine dimers from the DNA of
human cells in vitro, Biophys. J. 8, 319-325.

Regan, J.D., Setlow, R.B., and Ley, R.D. (1971) Normal and defective
repair of damaged DNA in human cells; a sensitive assay utilizing
the photolysis of bromodeoxyuridine, Proc. Nat. Acad. Sci. USA 68,
708-712.

Roth, H.D., and Lamola, A.A. (1972) Cleavage of thymine dimers
sensitized by quinones. Chemically induced dynamic nuclear polari-
zation in radical ions, J. Am. Chem. Soc. 94, 1013-1014.

Rupp, W., Wilde, C., Reno, D., and Howard-Flanders, P. (1971) Exchanges between DNA strands in ultraviolet irradiated E. coli, J.
Mol. Biol. 31, 291-304.

Setlow, R.B., and Carrier, W.L. (1964) The disappearance of thymine
dimers from DNA: an error-correcting mechanism, Proc. Nat. Acad.
Sci. USA 51, 226-231.

Slor, H., and Lev, T. (1971) Acid deoxyribonuclease activity in
purified calf thymus nuclei, Biochem. J. 123, 993-995.

Smerdon, M.J., Tisty, T.D., and Lieberman, M.W. (1978) Distribution
of ultraviolet-induced DNA repair synthesis in nuclease sensitive
and resistant regions of human chromatin, Biochemistry 17, 2377-
2386.

Söderhall, S., and Lindahl, T. (1975) Mammalian DNA ligase. Serological evidence for two separate enzymes, J. Biol. Chem. 250, 8438-8444.

Söderhall, S., and Lindahl, T. (1976) DNA ligases of eukaryotes, FEBS Letters 67, 1-8.

Strauss, B.S. (1976) Repair of DNA adducts produced by alkylation, in "Ageing, Carcinogenesis, and Radiation Biology" (K.C. Smith, ed.) pp. 287-314, Plenum Press, New York.

Sutherland, B.M., Runge, P., and Sutherland, J.C. (1974) DNA photoreactivating enzyme from placental mammals: origin and characteristics, Biochemistry 13, 4710-4714.

Takeshita, M., Grollman, A.P., Ohtsubo, E., and Ohtsubo, H. (1978) Interaction of bleomycin with DNA, Proc. Nat. Acad. Sci. USA 75, 5983-5987.

Talpaert-Borlé, M., Clerici, L., and Campagnari, F. (1979) Isolation and characterization of a uracil-DNA glycosylase from calf thymus, J. Biol. Chem. 254, 6387-6391.

Wang, E., and Furth, J.J. (1977) Mammalian endonuclease, DNase V. Purification and properties of enzyme of calf thymus, J. Biol. Chem. 252, 116-124.

Weiss, B. (1976) Endonuclease II of Escherichia coli is Exonuclease III*, J. Biol. Chem. 251, 1896-1901.

Yoneda, M., and Bollum, F. (1965) Deoxynucleotide-polymerizing enzymes of calf thymus glands. I. Large scale purification of terminal and replicative deoxynucleotide transferase, J. Biol. Chem. 240, 3385-3391.

Yoshida, S., Kondo, T., and Ando, T. (1974) Multiple molecular species of cytoplasmic DNA polymerase from calf thymus, Biochem. Biophys. Acta 353, 463-474.

IMMUNOREGULATORY T CELL SUBPOPULATIONS IN MAN

M. C. Mingari, A. Moretta, W. G. Canonica,* M. Colombatti
and L. Moretta

Istituto di Microbiologia del l'Università e I.S.M.I.*

Università di Genova, 16132 Genova

A large number of experimental evidence has been collected indicating that thymus derived (T) lymphocytes are by no means an homogeneous cell population. These cells mediate a spectrum of immune
responses such as delayed hypersensitivity, graft rejection, and
antineoplastic and antiparasitic immunity. In addition they play a
central role in humoral immunity by regulating B cell responses in
both positive (helper) and negative (suppressor) ways (Gershon, 1974).
In mice these various properties have been attributed to distinct
subclasses of T lymphocytes identified by different membrane markers.
The most promising marker for the identification of functional subsets of T cells has been the Ly antigen system. T cells bearing Ly 1
alloantigen show helper activity in B cell responses, respond in
mixed lymphocyte reactions to determinants coded for by the I and S
regions, and are the effector cells in delayed hypersensitivity reactions. Ly 2 and 3 alloantigens are present on suppressor T cells and
on CTL, and respond in MLR by recognizing K and D region products on
the allogeneic target cells (Cantor and Boyse, 1975 a, b; Cantor and
Boyse, 1977; Huber et al., 1976).

Recently the detection of Fc receptors for IgG on a minority of
normal mouse T lymphocytes provided another useful tool to dissect
murine T cells into two distinct subpopulations bearing or lacking
these markers (FcR$^+$ and FcR$^-$ cells respectively). The FcR$^-$ populations contain helper cells and the precursors of cytotoxic T lymphocytes (CTL). The FcR$^+$ cells include differentiated CTL and do not
have helper activity. Both FcR$^+$ and FcR$^-$ populations respond in MLR
(Stout and Herzenberg, 1975a, b; Stout et al., 1976).

Similar studies of human T cell subsets have been hampered by the lack of antisera specific to surface alloantigens and by obvious difficulties associated with human experimentation. A possibility to recognize different subpopulations of human T lymphocytes was offered by the observation that T cells from normal individuals have surface receptors for the Fc portion of IgG and by the discovery of a receptor for IgM (Moretta et al., 1975; Ferrarini et al., 1975). Both receptors for IgG and IgM were detectable only on a fraction of peripheral blood T lymphocytes, and a single T cell had one or the other receptor for immunoglobulins (Moretta et al., 1976). Thus our observation indicated that receptors for IgG and IgM represent useful markers for two different subpopulations of T cells which have been called T_M (or T_μ) and T_G (or T_γ) (Moretta et al., 1976).

Receptors for IgG and IgM have been detected by rosette formation with bovine erythrocytes coated with purified IgG and IgM antibodies. The use of this technique allows the isolation of rosette-forming cells on density gradients, and therefore the purification of T_M and T_G cells (Moretta et al., 1976; Moretta et al., 1978). Thus isolated cells can be studied <u>in vitro</u> for their properties. This approach for studying human T cell subpopulations, together with others which have been developed in these last two years, such as separation of T cells according to the presence of receptors for lectins (Hellström et al., 1976), the identification of membrane antigens by means of heteroantisera (Evans et al., 1977) or by sera from some patients with Juvenile rheumatoid arthritis (Strelkauskas et al., 1978), may allow a more accurate definition of human T cell subpopulations.

Morphology of T_M and T_G Lymphocytes

Morphological studies of purified T_G and T_M cells have revealed interesting differences in both light and electron microscopy examination. Cytocentrifuged preparations stained with the Giemsa dye showed that T_M cells have the characteristics of small lymphocytes, with a thin rim of cytoplasm and dense accumulations of chromatin in the nuclei. T_G lymphocytes appear larger with more abundant, weakly basophilic cytoplasm containing typical granules. These granules do not contain the enzyme markers of lysosomes present in macrophages and granulocytes (Grossi et al., 1978).

Electron microscopy examination showed that the cytoplasm of T_M cells appears empty of the common cytoplasm organelles, with the exception of monoribosomes. The cytoplasm of T_G cells is relatively rich in mitochondria, rough endoplasmic reticulum and Golgi. The granules mentioned above appear surrounded by a membrane and contain an electron-dense matrix (Grossi et al., 1978).

Distribution of T_M and T_G Cells in the Lymphoid Tissues

T_M and/or T_G cells are detectable in relevant proportions in all
the lymphoid organs with the exception of the thymus (Moretta et al.,
1978). Thus T_M cells are present in large proportions in peripheral
blood, tonsils, and lymph nodes. T_G cells outnumber T_M cells in the
spleen, but are found in much lower proportions in the other lymphoid
organs and are completely lacking in lymph nodes. It is of interest
that elevated proportions of T_G cells are present in the newborn.
This observation has been correlated with the suppressor function that
newborn T_G cells may exert on the mother's immune responses (Oldstone
et al., 1977).

Since lymphocytes present in the thymus are immature T cells,
our observations indicate that receptors for G or M immunoglobulins
are expressed at later stages of the T cell development process.

Variations in the expression of surface antigens during the
maturation of T lymphocytes are well known in the mouse. T cell precursors acquire most of T cell specific antigens such as GIX, θ, TL,
and Ly within the mouse thymus. Peripheral T cells lose GIX and
TL antigens and express a smaller amount of θ antigen. Moreover,
a portion of murine peripheral T cells selectively lose Ly antigens.
Therefore in the peripheral lymphoid organs, together with cells
which have all of the Ly alloantigens (Ly 1, 2, and 3), are present
cells which have selectively lost Ly 1 or Ly 2 and 3 antigens.

Functional Studies on T_M and T_G Subpopulations

Different _in vitro_ analyses of T cell function have been performed on purified T_M and T_G subpopulations. Thus, it has been
shown that T_M and T_G cells have a different responsiveness to PHA
but not to Con A.

The most interesting functional difference observed so far
regards their interaction with B lymphocytes for antibody responses.
In these studies we analyzed the effect of T cell subsets on B lymphocyte response to pokeweed mitogen (PWM). PWM induces polyclonal
lymphocyte proliferation and induces maturation of B lymphocytes
into antibody secreting cells. After PWM stimulation, a certain
proportion of recovered cells has the morphological characteristics
of plasma blasts and plasma cells. Immunoglobulins are detectable
in the cytoplasm of these cells by immunofluorescence on fixed preparations or are secreted in the supernatants in amounts measurable by
radioimmunoassay. Such B cell maturation, however, is dependent
upon the presence of T cells (Moretta et al., 1977; Moretta et al.,
1978). Thus the response of B lymphocytes to PWM is a suitable model
system for studies on regulatory mechanisms exerted by T lymphocytes

on B cell responses in humans. In order to identify the T cell sub-
population(s) which induces B cell proliferation and differentiation,
various T cell populations were cultured with B cells and PWM. Helper
function was measured after 7 days in culture by measuring the proliferative response and the numbers of cells with intracytoplasm immunoglobulins (cIg^+ cells). B cell maturation was induced only by
unfractionated T cells and T_M cells, whereas T_G cells or T cell populations depleted of T_M lymphocytes were ineffective in inducing B
cell proliferation and differentiation. These experiments clearly
indicate that cells with helper activity are restricted to T_M populations. Therefore T_G cells, as well as murine Ly 2 and 3 lymphocytes,
are incapable of helper cell function. We considered the possibility
that for analogy to the mouse system T_G lymphocytes contained cells
with suppressor activity. T_G lymphocytes, when added to a mixture
of T_M, B cells, and PWM, efficiently suppressed B cell differentiation. Whereas the helper activity of T_M lymphocytes is not affected
by their γ-irradiation (3.000 rads), suppressor activity is completely
abolished by irradiation of T_G cells prior to culture.

The suppressor capacity was the exclusive property of T_G lymphocytes since other cell populations, such as unfractionated T cells,
T_M, macrophages, and non-E-rosetting cells (therefore B cells plus
non-T-non-B-lymphocytes), were ineffective in inhibiting B cell responses to PWM.

In a further attempt to clarify the mechanism by which T_G cells
suppress B lymphocyte responses, we showed that T_G cells suppress
via soluble factors released without cell proliferation (lack of
tritiated thymidine uptake). In order to release suppressor factors
T_G cells require to be activated both by IgG immune complexes and by
PWM (antigen <u>in vivo</u>?). Suppression is observed only when T_G cells
are added early to mixtures of T helper and B cell cultured with PWM.

Another group of experiments provided good evidence about the
cell target of suppression. This could act either on helper T cells
or directly on B cells. In one series of experiments B cells were
"preactivated" by a previous coculture period with T_M cells and PWM.
These preactivated B cells continued to differentiate in culture
without need for further help. The addition of T_G cells to these
cultures did not suppress B cell differentiation. Thus, an early
helper signal provided to B lymphocytes cannot be reversed by a
subsequent suppressor signal from T_G cells. Further evidence that
B cells are not directly affected by suppressor signals was provided
by experiments in which T cell help was substituted by helper supernatants in inducing B cell differentiation as originally reported
by Janossy and Greaves (Janossy and Greaves, 1975). Under these
conditions T_G cells did not induce significant suppression of B
cell maturation.

All of the above results support the idea that suppressor factors released by T_G cells act on helper T cells or on their immediate precursors within the T_M population and not directly on B cells. This conclusion is supported by the observation that preincubation of T_M cells with suppressor supernatants sharply inhibited their capacity of promoting B cell differentiation. In view of the immunoregulatory role exerted by T_M and T_G subpopulations on antibody responses it is not surprising to detect important imbalances of these subpopulations in immunoglobulin deficiencies and autoimmune syndromes. Indeed imbalances of T cell subpopulations were consistently present in patients who had congenital or acquired abnormalities of the thymus, severe combined immunodeficiency, or defective T cell immunity without defined abnormalities of the thymus. Most of these patients had low numbers of circulating T_M cells and often increased proportions of T_G cells (Moretta et al., 1977b).

In patients with systemic lupus erythematosus, the observation of the immunoregulatory control resulting in an excessive production of immunoglobulins and of autoantibodies could be the consequence of a defective suppressor T cell activity. Indeed deficiencies of T_G cells are consistently present in patients with active disease (Moretta et al., 1979).

REFERENCES

Cantor, H., and E. A. Boyse, 1975a, Functional subclasses of T lymphocytes bearing different Ly antigens. I. The generation of functionally distinct T-cell subclasses is a differentiation process independent of antigen, J. Exp. Med., 141:1376.
Cantor, H., and E. A. Boyse, 1975b, Functional subclasses of T lymphocytes bearing different Ly antigens. II. Cooperation between subclasses of Ly[+] cells in the generation of killer activity, J. Exp. Med., 141:1390.
Cantor, H., and E. Boyse, 1977, Regulation of the immune responses by T-cell subclasses, in: "Contemporary Topics in Immunobiology," Vol. 7, O. Stutman, ed., pp. 47-67, Plenum Press, New York.
Evans, R. L., J. M. Breard, H. Lazarus, S. F. Schlossman, and L. Chess, 1977, Detection, isolation, and functional characterization of two human T-cell subclasses bearing unique differentiation antigens, J. Exp. Med., 145-221.
Ferrarini, M., L. Moretta, R. Abrile, and M. L. Durante, 1975, Receptors for IgG molecules on human lymphocytes forming spontaneous rosettes with sheep red cells, Eur. J. Immunol., 5:70.
Gershon, R. K., 1974, T cell control of antibody production, in: "Contemporary Topics in Immunobiology," Vol. 3, M. D. Cooper and N. L. Warner, eds.,pp. 1-40, Plenum Press, New York.

Grossi, C. E., S. R. Webb, A. Zicca, P. M. Lydyard, L. Moretta,
     M. C. Mingari, and M. D. Cooper, 1978, Morphological and histo-
     chemical analysis of two human T-cell subpopulations bearing
     receptors for IgM or IgG, J. Exp. Med., 147:1405.
Hellström, U., M. L. Dillner, S. Hammerström, and P. Perlmann, 1976,
     Fractionation of human T lymphocytes on wheat germ agglutinin-
     Sepharose, J. Exp. Med., 144:1381.
Huber, B., O. Devinsky, R. K. Gershon, and H. Cantor, 1976, Cell-
     mediated immunity: delayed-type hypersensitivity and cyto-
     toxic responses are mediated by different T-cell subclasses,
     J. Exp. Med., 143:1534.
Janossy, G., and M. Greaves, 1975, Functional analysis of murine
     and human B lymphocytes subsets, Transplant. Rev., 24:177.
Moretta, L., M. Ferrarini, M. L. Durante, and M. C. Mingari, 1975,
     Expression of a receptor for IgM by human T cells in vitro,
     Eur. J. Immunol., 5:565.
Moretta, L., M. Ferrarini, M. C. Mingari, A. Moretta, and S. R.
     Webb, 1976a, Subpopulations of human T cells identified by
     receptors for immunoglobulins and mitogen responsiveness,
     J. Immunol., 117:2171.
Moretta, L., S. R. Webb, C. E. Grossi, P. M. Lydyard, and M. D.
     Cooper, 1977a, Functional analysis of two human T-cell sub-
     populations: Help and suppression of B cell responses by T
     cells bearing receptor for IgM or IgG, J. Exp. Med., 146:184.
Moretta, L., M. Ferrarini, and M. D. Cooper, 1978b, Characterization
     of human T-cell subpopulations as defined by specific recep-
     tors for immunoglobulins, Contemp. Top. Immunobiol, 8:19.
Oldstone, M. B. A., A. Tishon, and L. Moretta, 1977, Active thymus
     derived suppressor lymphocytes in human cord blood, Nature,
     269:333.
Strelhauskas, A. J., V. Shauf, B. S. Wilson, L. Chess, and S. F.
     Schlossman, 1978, Isolation and characterization of naturally
     occurring subclasses of human peripheral blood T cells with
     regulatory functions, J. Immunol., 120:1278.

THYMOCYTE MATURATION INDUCED BY A CYCLIC

AMP-ELEVATING THYMIC FACTOR

A. Astaldi and G.C.B. Astaldi, Central Laboratory of the

Netherlands Red Cross Blood Transfusion Service, P.O.Box

9190, Amsterdam, The Netherlands

INTRODUCTION

The differentiation of T lymphocytes into mature immunocompetent cells is probably at least partially under the influence of
factors secreted by the thymus, as suggested by experiments in a
diffusion chamber (1,2). Several factors, acting on T-cell differentiation, have been found in thymic extracts, in peripheral blood
and in supernatants of thymic epithelial cultures (for review, see
ref. 3). Probably, these factors act via specific membrane receptors. Indirect evidence (4) suggests that cyclic AMP (cAMP) may
play a 'second messenger' role in the induction of T-cell differentiation by thymic factors. A thymic humoral factor (THF) (5) and a
thymic epithelial culture supernatant (TES) (6) also induce an increase in intracellular cAMP levels in thymocytes. By means of
direct measurement of intracellular cAMP in human and mouse thymocytes in vitro, we demonstrated the existence in human serum of a
thymus-dependent factor (SF) (7) which has the capacity to increase
cAMP levels.

The factor

The serum factor SF was isolated from human blood (7,8) and
found to be thymus-dependent because SF appeared to be absent in
thymectomized human donors; after thymectomy, the SF activity decreased rapidly (7,9). When the activity of SF was measured in
healthy donors from birth to 90 years of age, it was found to be
present in the first 10 years of life, reaching its maximum in the
3rd decade and decreasing thereafter (10,11), in analogy with several other immune functions (12). SF was also found to be absent in
all patients with thymus-dependent immunodeficiency diseases tested

so far. When the thymic extract thymosin was injected into patients
lacking SF, it induced the appearance of SF, or SF-like material
(13,14).

SF is distinct from FTS, a thymus-dependent peptide isolated
by Bach et al. (15) from pig serum, because FTS has different biological activities (16) and biochemical properties as compared to
SF.

Target cell specificity

SF was found to act markedly on mouse thymocytes, moderately
on lymph node cells and only marginally on spleen cells (7). No
activity of SF was found on other lymphoid and non-lymphoid cells
(7,8,17). These findings, indicating that most target cells for SF
are among thymocytes, prompted us to investigate to which subpopulation of thymocytes do target cells for SF belong. Mouse thymocytes can be distinguished in two subpopulations: one mainly present
in the cortex, sensitive to the lytic action of cortisone, can be
agglutinated by the lectin peanut agglutinin and is immunologically
immature; the other, resistant to cortisone, cannot be agglutinated
by the peanut lectin and is mainly present in the medulla (18).
This latter subpopulation was reported to be as immunocompetent as
mature T cells (18,19). These two subpopulations of thymocytes
were isolated by means of cortisone administration in vivo (8) to
obtain the immunocompetent cortisone-resistant one and by means of
agglutination with peanut lectin (18) to obtain the immunologically
immature, cortisone-sensitive population. SF was found to act only
on the cortisone-sensitive cells; no activity of SF was observed on
cortisone-resistant cells. We concluded that target cells for SF
are among immunologically immature thymocytes, thus among post-
thymic precursor cells.

Early events induced by SF in thymocytes

In 1976, we reported (7) that SF increases intracellular
cyclic AMP (cAMP) levels in thymocytes. Because SF does not induce
any change in the level of intracellular cyclic GMP (cGMP), we concluded that SF alters the cAMP/cGMP ratio in favour of cAMP, an
event associated with cell differentiation (20). The effect of SF
on the level of cAMP, evident after 1 min of incubation at $37^{o}C$,
attains a peak after 5 min and decreases thereafter (21). This
type of kinetic rules out the possibility that SF might increase
cellular cAMP by interfering with the degradation process of cAMP,
thus indicating that SF does not act as an inhibitor of phosphodi-
esterases and suggesting a direct action of SF on the production of
cAMP. Therefore, we tested the effect of SF on thymocyte membranes
and found that indeed SF stimulates adenylate cyclase to produce
cAMP. SF was also found (7) to be selectively adsorbed by thymocytes and to be not inhibited by propranolol, a β-adrenergic antago-
nist. We concluded (7,9) that, most likely, SF acts through a mem-

brane site, probably a receptor, distinct from that for β-adrenergic activators.

Using an indirect approach, Storrie et al. (22) suggested that
protein synthesis is one of the biochemical events induced by thymic factors. Indeed, we found (21) that SF stimulates the incorporation of ^{3}H-leucine into proteins of thymocytes cultured for 2-4
h. Because the activity of SF is completely blocked by cyclohexi-
mide, a specific inhibitor of protein synthesis, we concluded that
SF stimulates protein synthesis in thymocytes. It is known that
for cell maturation synthesis of nuclear proteins is required (23).
Several pieces of evidence (24,25) indicate that, among the nuclear
proteins, the group of non-histone chromosomal proteins (NHCP) contains macromolecules that are involved in the regulation of chromatin template activity. These findings prompted us to perform qualitative studies on the proteins synthesized under the influence of
SF. It was found (26) that SF stimulates the incorporation of ^{3}H-
leucine only in nuclear proteins. When the distribution of radioactivity in the various nuclear proteins (phosphorylated NHCP,
P-NHCP; histones; residual NHCP, R-NHCP) was compared, the incorporation of ^{3}H-leucine was found to be especially located in the P-
NHCP which represents only about 10% of the total nuclear proteins. At the level of histones, which represent the highest amount
of proteins recovered from the chromatin (about 40%), the radioactivity per mg protein was one-tenth of that found in the P-NHCP
fraction. SF also induced incorporation of ^{32}P selectively into
P-NHCP. Electrophoretic analysis of P-NHCP showed that bands of
proteins, with molecular weight higher than 50×10^3 daltons, are
synthesized to a larger extent in thymocytes stimulated with SF for
4 h. We concluded that SF stimulates synthesis and phosphorylation
of high molecular weight P-NHCP.

Because SF does not affect the number of proliferating thymocytes (27) and does not induce changes at the histone level (26),
the activity of SF on thymocytes is compatible with DNA translation
and transcription, but not with DNA replication.

We found similar biochemical events to be induced also by
other cAMP-elevating agents, such as the β-adrenergic stimulator
isoproterenol and by prostaglandin E_1 (PGE_1) (7,8,17,21,26).

Effect of SF on thymocyte maturation

As reported above, target cells for SF are among immunologically immature thymocytes. Blomgren and Andersson (19) demonstrated that immature T cells are sensitive to hydrocortisone. Trainin
et al. (28) reported that the thymic humoral factor THF increases
the resistance of thymocytes to hydrocortisone treatment _in vitro_
and suggested that this might be regarded as a maturation process.
Incubation of thymocytes with SF enlarged the population of hydrocortisone-resistant cells (8), as reported for THF (28). We also
found (8) that the cAMP-increasing agent isoproterenol induces
hydrocortisone resistance similar to SF.

We concluded that the induction of hydrocortisone resistance might be associated with the increase in cAMP, in agreement with data (4,29), demonstrating that the differentiation pathway of committed precursors to mature T cells involves an increase in cellular levels of cAMP and that such differentiation can be induced both by thymic and non-thymic cAMP-elevating agents.

Immunologically mature human T cells have been shown to bear Fc receptors for IgG (Tγ) and for IgM (Tμ). Functional analysis has indicated a suppressor activity for Tγ and a helper activity for Tμ (30). Investigations on tissue distribution of Tγ and Tμ revealed the presence of at least one of these two subpopulations in all lymphoid tissues except the thymus (31). Incubation with SF induced the capability to acquire the Fc receptor for IgM in about 9% of the thymocytes (17). No effect of SF could be observed on the percentage of cells expressing the Fc receptor for IgG. We concluded that SF induces also some phenotypic traits of immuno-competent human T cells.

Terminal deoxynucleotidyl transferase (TdT), a DNA-polymerizing enzyme, which adds deoxyribonucleotides to an DNA primer in the absence of a template, was first described by Bollum (32). Under normal conditions, TdT is found only in precursors of immuno-competent T cells, such as hydrocortisone-sensitive thymocytes and bone marrow cells, and is absent from fully immunocompetent peripheral T cells (for review, see refs. 33 and 34). These findings indicate that the final maturation of T cells is associated with the disappearance of TdT activity and have led Baltimore (35) to suggest that TdT is a potential somatic mutagen, possibly involved in the generation of immunological diversity. SF induces a marked decrease of the TdT activity both in human and in mouse thymocytes (36) after 240 min of exposure. We concluded that, if TdT is in-volved in the generation of immunological diversity, as proposed by Baltimore (35), SF might act either on fully diversified cells and so induce their maturation or on cells potentially capable of further diversification. Preliminary indications suggest that the second possibility might be correct.

A graft-versus-host (GvH) reaction occurs when the grafted cells are immunocompetent to react against the host, but the host is not immunocompetent to reject the graft. A typical situation of GvH reaction is achieved by the injection of homozygous parental (P) lymphoid cells into an F_1 hybrid mouse. Simonsen (37) has developed a method for the bio-assay of immunological activity in suspensions of lymhoid cells. The method is based on the finding that the degree of splenomegaly, which develops in the early stages of the GvH reaction, is correlated with the number of immunologi-cally competent cells grafted into the host. Using this method, it was shown (19) that cortisone-sensitive thymocytes are immunologi-cally immature. This was recently confirmed, using a similar method, by Irlè et al. (18), who showed that cortisone-sensitive thymocytes can be agglutinated by the peanut lectin and are immunologically immature.

After incubation with SF, peanut lectin-agglutinating, cortisone-sensitive thymocytes were found to be more effective in inducing the GvH reaction; furthermore, treatment with SF had no effect
on the capacity of cortisone-resistant thymocytes to induce the GvH
reaction (27). We concluded that at least part of the cortisone-
sensitive, immunologically immature thymocyte population can
acquire properties of functionally immunocompetent cells under the
influence of SF.

CONCLUSION

The stimulation of adenylate cyclase is likely to be the
signal, delivered by SF to post-thymic T-cell precursors, which
initiates their maturation to immunocompetent T cells. In fact,
elevated cAMP levels, as a result of stimulation of adenylate cyclase, probably enhance the availability of phosphate groups and
stimulate protein kinases, as suggested by the experiments on the
incorporation of ^{32}P into nuclear proteins of thymocytes (26).
Protein kinases are known to induce phosphorylation and synthesis
of nuclear proteins, as we found to happen after stimulation of
adenylate cyclase in thymocytes (26). The newly formed phosphorylated nuclear proteins would then be responsible for gene activation (38) which would result into the phenotypic expression of
genetic information. In immature thymocytes, the result of gene
activation should be maturation to immunocompetent cells. This was
shown to be caused by SF, because SF induces in thymocytes several
properties of mature T cells, such as resistance to hydrocortisone
(8), low terminal deoxynucleotidyl transferase activity (36) and
capacity to elicit an effective graft-versus-host reaction (27).

If the elevated cAMP level (or the elevated cAMP/cGMP ratio)
would indeed be the signal which turns on thymocyte maturation, one
would expect that any agent increasing cAMP in thymocytes induces
their maturation. This was indeed shown to be the case by several
experimental data (4,8,26,29,39).

In this case, we should face the problem of how does a thymocyte recognize a specific factor (e.g. SF) from a non-specific
factor (e.g. isoproterenol) if they do share the same signal, i.e.
increase in cAMP. Inasmuch as phenomena, such as cAMP compartmen-
talization, have been largely quoted but never proven, it seems
likely that the recognition phase occurs at the membrane level.

We showed (7,8,17) that "receptors" for SF are present only on
immature T cells. On the contrary, receptors for non-specific
stimulants (e.g. isoproterenol and PGE$_1$) are known to be present on
several cell types, including mature T cells, B cells and non-
lymphoid cells (7,8,17,40,41,42). The concentration of such stimulants, necessary to elevate cAMP levels in post-thymic T-cell
precursors, is not only higher than the one of SF, but also much
higher than the concentration normally reached by such agents _in_
vivo. As a consequence, under normal circumstances, non-specific

stimulants are unlikely to induce T-cell maturation, because they are present in too low concentrations to act on precursors of mature T cells: at these physiological concentrations, the activity of such stimulants is restricted to their physiological target cells (e.g. intestinal smooth muscle for β-adrenergic activators).

As mentioned, non-specific stimulants can be used in high concentrations to induce T-cell maturation in vitro (8,26,29,39), but it should be kept in mind that this is an artificial situation which does not reflect the one in vivo: when used at such high concentrations in vivo, non-specific stimulants are likely to activate a very large number of cell types, probably leading to paralysis.

REFERENCES

1. Stutman, O. and Good, R.A. Thymus hormones. In: Contemporary Topics in Immunobiology. A.J.S. Davies and R.L. Carter, editors. Plenum Press, New York, Vol. 2, 299 (1973).
2. Stutman, O., Yunis, E.J., Martinez, C. and Good, R.A. Reversal of postthymectomy wasting disease in mice by multiple thymus grafts. J. Immunol. 98, 79 (1967).
3. Khan, A. Thymic hormones and immunopeptides. Ann. Allergy 41, 78 (1978).
4. Scheid, H.P., Goldstein, G., Hämmerling, V. and Boyse, E.A. Lymphocyte differentiation from precursor cell in vitro. Ann. N.Y. Acad. Sci. (USA) 249, 531 (1975).
5. Kook, A.I. and Trainin, N. Hormone-like activity of the thymic humoral factor on the induction of immune competence in lymphoid cells. J. Exp. Med. 139, 193 (1974).
6. Kruisbeek, A.M., Astaldi, G.C.B., Blankwater, M.J., Zijlstra, J.L., Levert, L.A. and Astaldi, A. The in vitro effect of a thymic epithelial culture supernatant on mixed lymphocyte reactivity and intracellular cAMP levels of thymocytes and on antibody production to SRBC by Nu/Nu spleen cells. Cell. Immunol. 35, 134 (1978).
7. Astaldi, A., Astaldi, G.C.B., Schellekens, P.Th.A. and Eijsvoogel, V.P. Thymic factor in human sera demonstrable by a cyclic AMP assay. Nature 260, 713 (1976).
8. Astaldi, G.C.B., Astaldi, A., Groenewoud, M., Wijermans, P., Schellekens, P.Th.A. and Eijsvoogel, V.P. Effect of a human serum thymic factor on hydrocortisone-treated thymocytes. Eur. J. Immunol. 7, 836 (1977).
9. Astaldi, A., Astaldi, G.C.B., Schellekens, P.Th.A. and Eijsvoogel, V.P. Is there a circulating human thymic factor that induces cyclic AMP synthesis? Reply. Nature 271, 666 (1978).
10. Wijermans, P. and Astaldi, A. Effect of aging on thymus-dependent serum factor(s). Ned. T. Geront. 9, 216 (1978).
11. Wijermans, P., Oosterhuis, H.J.G.H., Astaldi, G.C.B.,

Schellekens, P.Th.A. and Astaldi, A. Thymus-dependent serum factor in non-thymectomized and thymectomized myasthenia gravis patients. Submitted for publication.

12. Wijermans, P., Oosterhuis, H.J.G.H., Astaldi, G.C.B., Schellekens, P.Th.A. and Astaldi, A. Influence of adult thymectomy on the immunocompetence in man. Submitted for publication.

13. Astaldi, A., Astaldi, G.C.B., Wijermans, P., Groenewoud, M., Schellekens, P.Th.A. and Eijsvoogel, V.P. Thymosin-induced human serum factor increasing cyclic AMP. J. Immunol. 119, 1106 (1977).

14. Astaldi, A., Astaldi, G.C.B., Wijermans, P., Dagna-Bricarelli, F., Kater, L., Stoop, J.W. and Vossen, J.M. Experiences with thymosin in primary immunodeficiency disease. Cancer Treat. Rep. 62, 1779 (1978).

15. Bach, J.F., Dardenne, M., Pleau, J.M. and Rosa, J. Biochemical characterization of a serum-thymic factor. Nature 266, 55 (1977).

16. Bach, M.A. Sera from both normal and thymectomized animals produce elevations of cyclic AMP in thymocytes. Cell. Immunol. 33, 224 (1977).

17. Astaldi, A., Astaldi, G.C.B., Wijermans, P., Groenewoud, M., Bemmel, T. van, Schellekens, P.Th.A. and Eijsvoogel, V.P. Thymus-dependent human serum factor active on precursors of mature T cells. In: Proc. 12th Leukocyte Culture Conference. M. Quastel, editor. Academic Press, Inc., New York, in press.

18. Irlè, C., Piguet, P.-F. and Vassalli, P. In vitro maturation of immature thymocytes into immunocompetent T cells in the absence of direct thymic influence. J. Exp. Med. 148, 32 (1978).

19. Blomgren, H. and Andersson, B. Evidence for a small pool of immunocompetent cells in the mouse thymus. Exp. Cell Res. 57, 185 (1969).

20. Watson, J. The influence of intracellular levels of cyclic nucleotides on cell proliferation and the induction of antibody response. J. Exp. Med. 141, 97 (1975).

21. Wijermans, P., Astaldi, G.C.B., Facchini, A., Schellekens, P.Th.A. and Astaldi, A. Early events in thymocyte activation. I. Stimulation of protein synthesis by a thymus-dependent human serum factor. Biochem. Biophys. Res. Commun. 86, 88 (1979).

22. Storrie, B., Goldstein, G., Boyse, E.A. and Hämmerling, V. Differentiation of thymocytes: Evidence that induction of the surface phenotype requires transcription and translation. J. Immunol. 116, 1358 (1976).

23. Kishimoto, T., Nishizawa, Y., Kikutani, H. and Yamamura, Y. Biphasic effect of cyclic AMP on IgG production and on the changes of non-histone nuclear proteins induced with anti-immunoglobulin and enhancing soluble factor. J. Immunol.

$\underline{118}$, 2027 (1977).

24. Stein, G.S., Stein, J.L., Kleinsmith, L.J., Thompson, J.A., Park, W.D. and Jansing, R.L. Role of non-histone chromosomal proteins in the regulation of histone gene expression. Cancer Res. $\underline{36}$, 4307 (1976).

25. Elgin, S.C.R. and Weintraub, H. Chromosomal protein and chromatin structure. Ann. Rev. Biochem. $\underline{44}$, 725 (1975).

26. Facchini, A., Astaldi, G.C.B., Cocco, L., Wijermans, P., Manzoli, F.A. and Astaldi, A. Early events in thymocyte activation. II. Changes in non-histone chromatin proteins induced by a thymus-dependent human serum factor. Submitted for publication.

27. Astaldi, G.C.B., Astaldi, A., Wijermans, P., Schellekens, P.Th.A. and Eijsvoogel, V.P. A thymus-dependent serum factor induces maturation of thymocytes as evaluated by Graft-versus-Host reaction. Submitted for publication.

28. Trainin, N., Levo, Y. and Rotter, V. Resistance to hydrocortisone conferred upon thymocytes by a thymic humoral factor. Eur. J. Immunol. $\underline{4}$, 634 (1974).

29. Horowitz, S.H. and Goldstein, A.L. The in vitro induction of differentiation potative human stem cells by thymosin and agents that affect cyclic AMP. Clin. Immunol. Immunopathol. $\underline{9}$, 408 (1978).

30. Moretta, L., Webb, S.R., Grossi, C.E., Lydyard, P.M. and Cooper, H.D. Functional analysis of two human T-cell subpopulations: help and suppression of B-cell responses by T cells bearing receptors for IgM or IgG. J. Exp. Med. $\underline{146}$, 184 (1977).

31. Moretta, L., Ferrarini, M. and Cooper, H.D. Characterization of human T-cell subpopulations as defined by specific receptors for immunoglobulins. Contemp. Top. Immunobiol. $\underline{8}$, 19 (1978).

32. Bollum, F.J. Calf thymus polymerase. J. Biol. Chem. $\underline{235}$, 2399 (1960).

33. Bollum, F.J. Terminal deoxynucleotidyl transferase. In: The Enzymes. R.D. Boyer, editor. Academic Press, Inc., New York, Vol. 10, pp. 145-171 (1974).

34. Baltimore, D., Silverstone, A.E., Kung, P.C., Harrison, T.A. and McCaffrey, R.P. What cells contain terminal transferase? In: The Generation of Antibody Diversity: a new look. A. Cunningham, editor. Academic Press, Inc., New York, pp. 21-30 (1977).

35. Baltimore, D. Is terminal deoxynucleotidyl transferase a somatic mutagen in lymphocytes? Nature $\underline{248}$, 409 (1974).

36. Astaldi, G.C.B., Astaldi, A., Wijermans, P., Bemmel, T. van, Schellekens, P.Th.A. and Eijsvoogel, V.P. A thymus-dependent human serum factor induces decrease of terminal deoxynucleotidyl transferase in thymocytes. Submitted for publication.

37. Simonsen, M. Graft-versus-host reactions. Their natural
     history and applicability as tools of research. Progr.
     Allergy 6, 349 (1962).
38. Stein, G.S., Chaudmury, S.C. and Baserga, R. Gene activation
     in WI-38 fibroblasts stimulated to proliferate. Role of
     non-histone chromosomal proteins. J. Biol. Chem. 247, 3918
     (1972).
39. Scheid, H.P., Hoffmann, H.K., Komura, K., Hämmerling, V.,
     Abbot, J., Boyse, E.A., Cohen, G.H., Hooper, J.A., Schulof,
     R.S. and Goldstein, A.L. Differentiation of T cells induced
     by preparations from thymus and by non-thymic agents. J. Exp.
     Med. 138, 1027 (1973).
40. Smith, J.W., Steiner, A.L., Newberry, H.W. and Parker, C.W.
     Cyclic adenosine 3',5'-monophosphate in human lymphocytes.
     Alterations after phytohemagglutinin stimulation. J. Clin.
     Invest. 50, 432 (1971).
41. Bach, M.A. Differences in cyclic AMP changes after stimula-
     tion by prostaglandins and isoproterenol in lymphocyte sub-
     populations. J. Clin. Invest. 55, 1074 (1975).
42. Niaudet, P., Beaurain, G. and Bach, M.A. Differences in
     effect of isoproterenol stimulation on levels of cyclic AMP
     in human B and T lymphocytes. Eur. J. Immunol. 6, 834 (1976).

RECEPTORS AND CELL COLLABORATION IN THE IMMUNE SYSTEM

Klaus Rajewsky

Institute for Genetics, University of Cologne

Weyertal 121, D-5000 Köln 41, F.R.G.

1. INTRODUCTION

The immune system consists of a large number of lymphocytes which carry on the cell surface receptor molecules which specifically bind antigen. Each cell is committed to a single receptor specificity, and since the number of different receptor specificities in the immune system must be in the order of millions or more, most lymphocytes differ from each other in terms of receptor specificity. In addition to this specialization, lymphocytes mature in the bone marrow and are the precursors of antibody forming cells. T lymphocytes also originate in the bone marrow but differentiate in the thymus. They do not secrete antibodies and can be subdivided into many subpopulations each of which is characterized by a distinct function and a distinct phenotype of the cell surface. Killer T cells directly attack and lyse their cellular targets. Helper T cells enhance killer and B cell activities, suppressor T cells suppress them. Helper and suppressor cells thus regulate the effector functions of the immune system, and these regulatory cell interactions are specific in the sense that they occur specifically for a given immune response (for review see 1, 2).

It is clear that this specificity must have its basis in the antigen receptor molecules with which lymphocytes are equipped. These receptors thus mediate both antigen recognition by the immune system and the regulatory cell interactions by which the immune response is regulated.

In this short summary I shall concentrate on two points. The first is the nature of antigen receptors. The second is the problem of how receptor molecules mediate regulatory cell interactions.

2. ANTIGEN RECEPTORS: V REGIONS OF IMMUNOGLOBULIN H CHAINS ARE EXPRESSED IN T CELL RECEPTORS FOR ANTIGEN

The classical receptor in the immune system is the antibody molecule. It consists of two types of chains (L and H) each of which is composed of so-called constant (C) and variable (V) domains. The C domains repeat from one antibody molecule of a given class to another. The V domains are specific for individual antibody molecules and carry the antigen binding site. It is well established since many years that cells carry on their surface antibody molecules as receptors for antigen since antigen is bound to the B cell surface by receptors which also react with antisera specific for antibody C domains (3).

More recently antisera against V domains have also been developed. One class of such reagents reacts with determinants common to many or all V regions of a given group. There are three groups of V regions, namely those common to H chains (V_H) and the V regions of the two L chain types κ and λ (V_κ and V_λ). So far antisera against common determinants of V_H (anti-V_H) and V_λ (anti-V_λ) have been successfully raised (4, 5). A second class of V region-specific reagents are so-called anti-idiotypic antisera (6). Such sera are specific for a single antibody, i.e. for a particular pair of V domains. One says that the anti-idiotypic antibody or anti-idiotype recognizes the idiotype of the antibody molecule. Idiotypes are particularly valuable markers of antibody V domains. It has been shown that in many immune responses of inbred laboratory animals (e.g. mice) to simple antigens antibodies with a particular idiotype are reproducibly synthesized by most individuals. Such idiotypes are called "major" or "cross-reactive" idiotypes. Often the same idiotype is not produced in a different strain, and genetic studies have established simple Mendelian inheritance of major idiotypes. Idiotypes can thus be used as genetic markers of antibody V domains. In most cases a close linkage between idiotype and heavy chain allotype was found. Such idiotypes are thus in all likelihood markers of V domains of the heavy chain (for review see 7).

Antisera against V domains easily detect these structures on the surfaces of B cells, in accord with the view that B cells possess antibody-like receptors for antigen. How is the situation in the case of T lymphocytes?

The molecular nature of the T cell receptor for antigen is an issue of debate since long and the problem is still unresolved. In functional terms T cells appear to discriminate among antigens as well as B cells and antibodies, but in addition they have a tendency to react with structures encoded by the major histocompatibility complex (MHC). In fact, there is evidence that at least certain classes of T cells (like killer and helper cells) see antigen always in association with MHC products (reviewed in 8). This has entertained the view that T lymphocytes may have two kinds of receptors, one directed against MHC products and the other specific for conventional antigenic determinants. Be this as it may, the question remains which structures T cells make use of to discriminate among antigens including those of the MHC. More specifically we might ask whether T cells express in their receptors antibody V regions in the same manner as B cells.

In the absence of structural data serological methods have been used in order to answer this question, and the most convincing results have been obtained at the level of idiotypic analysis. Anti-idiotypic antisera have revealed the presence of idiotypic molecules on the T cell surface and such molecules have been isolated from T cells (9, 10). The specificity of the reaction was confirmed by genetic studies which also showed that the idiotypic structures on T cells are encoded by genes in the heavy chain linkage group as it would be expected for genes encoding V regions of the H chain. In addition, it was shown that the idiotypic molecules on T cells are a product of these cells (9, 11, 12) and that the cells expressing a given idiotype are functionally endowed with the corresponding antigen specificity (9) and the same was found to be true for isolated idiotypic T cell receptor molecules (9, 10). Finally, functional studies demonstrated that anti-idiotypic antibodies were able to specifically sensitize idiotype-bearing T cells in a manner analogous to sensitization by antigen (13).   Sensitivity to sensitization by anti-idiotype was found in genetic experiments to be linked to the heavy chain linkage group, again suggesting that T cell idiotypes are encoded by V_H genes (14). In accordance with these results is the finding that anti-V_H sera (see above) react with isolated T cell receptors (15) and also, as shown recently by Tada et al., with T cell derived, antigen-specific suppressor factors (T. Tada, personal communication).

The evidence thus strongly suggests that V regions of immunoglobulin H chains are part of the T cell receptor for antigen, and that there is thus idiotype sharing between receptors of T and B lymphocytes. A large number of questions remain unresolved, however. First, idiotypic crossreactivity between T and B cell receptors does not imply that the sets of V_H domains expressed in the two cellular compartments are identical; in fact, strong evidence indicates that they are not and that the rules of receptor diversification in T and B cells are different (15, 16). This finding is hardly surprising in view of the functional differences between the two cell types as outlined above. A second problem relates to the failure of several workers to detect V regions of L chain in T cell receptors (9, 17). Since antibody specificity is determined by the combination of V_H and V_L domains and T cell receptors at least in certain cases resemble antibodies in terms of affinity and fine specificity of antigen binding (10, 15), the question arises whether V_L regions still remain to be discovered

in T cell receptors or a light chain analogue substitutes for them.
Such an analogue could be encoded in the MHC and in particular the
I region of the MHC, since I region determinants have been identified in T cell-derived antigen-specific "factors" and in addition
genes in the I region (the so-called Ir genes) control T cell specificities in certain instances (for review see 8).

3. LYMPHOCYTE COOPERATION: THE IDIOTYPIC NETWORK

As pointed out in the introduction, immunological reactions are
usually the result of a series of specific lymphocyte interaction.
Regulator (helper or suppressor) cells control the activity of effector cells. The interactions are specific for a specific immune response and must therefore be mediated by specific receptor molecules.
How can this be achieved? The classical model of lymphocyte interactions implies that the interacting cells are specifically brought
together by the multideterminant antigen which binds specifically
to the receptors of the cells: The antigen bridge model (18). This
simplistic view has been considerably broadened by the concept of
idiotypes and anti-idiotypes. Cogent experiments have established
that idiotypic and anti-idiotypic receptors and antibodies coexist
in the immune system (19, reviewed in 20). The system is so diverse
that each antibody or receptor molecule does not only carry its own
idiotype but is at the same time an anti-idiotype against another
idiotype in the same system. In addition, both T and B lymphocytes
participate in the network since their receptors share idiotypes,
as mentioned above. The network hypothesis of Jerne views the immune
system as a network of cells which interact with each other directly
or indirectly via their idiotypic and anti-idiotypic receptors (21).
Formally, these kinds of interactions resemble antigen-mediated cell
interactions since the interaction of an anti-idiotype as well as
that of antigen with an antibodiy or receptor molecule takes place
in the V region of the latter molecule. The various kinds of recep-
tor-mediated cell interactions are formally represented in Fig. 1.

A large body of experimental evidence suggests that idiotypic
regulation can actually occur in the immune system. Anti-idiotypic
antibodies can sensitize T helper, T suppressor, T killer and B
cells (13, 22, 23). Anti-idiotypic helper cells help idiotypic B
cells to differentiate into antibody secreting cells (24, 25). T
cells and B cells with crossreacting idiotypes on their receptors
can interact via an anti-idiotypic (instead of an antigen) bridge
(26). In certain immune responses the production of antibody is
followed by the production of autologous anti-idiotypic antibody
(19). It is thus clear that the formal network can become functional
if the system is appropriately manipulated. Such manipulations may
become important in practical terms in the future. To which extent
their result reflects physiological regulatory processes in the
immune system remains to be established.

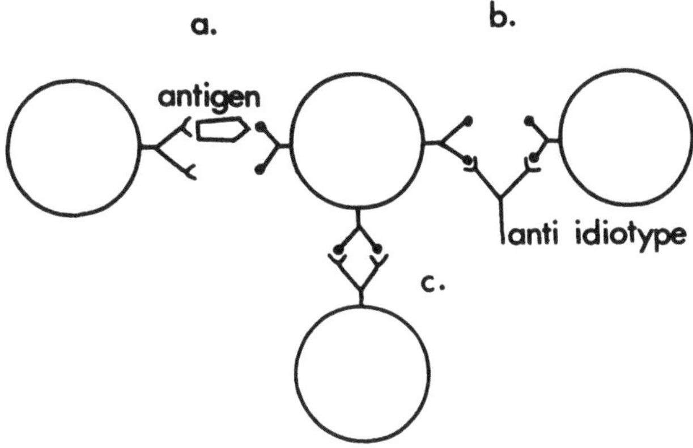

Models of lymphocyte interactions via a) an antigen bridge b) an
anti-idiotypic bridge and c) by direct receptor interaction.

4. SUMMARY

Specific immunological reactions are mediated by receptor molecules on the lymphocyte surface and by humoral antibodies. The receptors of B lymphocytes (the precursors of antibody forming cells)
are antibodies and carry V regions of immunoglobulin H and L chains.
T cell receptors express V regions of immunoglobulin H chains. It is
unresolved whether additional molecular structures participate in
the construction of the antigen binding site of these molecules. Since
both T and B cells express V_H regions in their receptors they are both
partners of a single network in which the cells interact with each
other through direct or indirect (antibody-mediated) receptor interactions. These kinds of cell interactions can be experimentally
demonstrated and may represent, together with the classical antigen-
mediated cell interactions, a major regulatory principle of the immune
system.

5. REFERENCES

1. D.H. Katz and B. Benacerraf, Adv. Immunol. 15: 2 (1972).
2. H. Cantor and E.A. Boyse, Cold Spring Harbor Symp. Quant. Biol.
   41: 23 (1977).
3. R.M.E. Parkhouse and E.R. Abney. In F. Loor and G.E. Roelants
   (eds.), B and T Cell in Immune Recognition, p. 211. J. Wiley and
   Sons, London, 1977.

4. Y. Ben-Neriah, P. Lonai, M. Gavish and D. Givol, Eur. J. Immunol.
   8: 792 (1978).
5. Y. Ben-Neriah, C. Wuilmart, P. Lonai and D. Givol, Eur. J. Immunol.
   8: 797 (1978).
6. J. Oudin and M. Michel, C.R. Acad. Sci. (Paris) 257: 805 (1963).
7. K. Eichmann, Immunogenetics 2:491 (1975).
8. K.F. Lindahl and K. Rajewsky, In E.S. Lennox (ed.), International
   Review of Biochemistry. Defense and Recognition IIA, Vol. 22,
   p. 97. University Park Press, Baltimore, 1979
9. H. Binz and H. Wigzell, Contemp. Top. Immunobiol. 7: 113 (1977).
10. U. Krawinkel, M. Cramer, T. Imanishi-Kari, R.S. Jack, K. Rajewsky
    and O. Mäkelä, Eur. J. Immunol. 8: 566 (1977).
11. M.H. Julius, H.C. Cosenza and A.A. Augustin, Eur. J. Immunol.
    8: 484 (1978).
12. U. Krawinkel, M. Cramer, B. Kindred and K. Rajewsky, Eur. J.
    Immunol. in press.
13. K. Eichmann and K. Rajewsky, Eur. J. Immunol. 5: 661 (1975).
14. G.J. Hämmerling, S.J. Black, C. Berek,, K. Eichmann and K. Rajewsky,
    J. Exp. Med. 143: 861 (1976).
15. M. Cramer, U. Krawinkel, I. Melchers, T. Imanishi-Kari, Y. Ben-
    Neriah, D. Givol and K. Rajewsky, Eur. J. Immunol. in press.
16. U. Krawinkel, M. Cramer, I. Melchers, T. Imanishi-Kari and K.
    Rajewsky, J. Exp. Med. 147: 1341 (1978).
17. K. Eichmann, In E.E. Sercarz, L.A. Herzenberg and C.F. Fox (eds.),
    Regulatory Genetics of the Immune System: ICN-UCLA Symposia on
    Molecular and Cellular Biology, Vol. VI, p. 127. Academic Press,
    New York, 1977.
18. N.A. Mitchison, K. Rajewsky and R.B. Taylor, In J. Sterzl and I.
    Riha (eds.), Prague Symposium on Developmental Aspects of Anti-
    body Formation and Structure, Vol. II, p. 547. Publishing House
    of the Czechoslovak Academy of Sciences, Prague, 1970.
19. H. Cosenza, Eur. J. Immunol. 6: 114 (1975).
20. K. Eichmann, Adv. Immunol. 26: 195 (1978).
21. N.K. Jerne, Ann. Immunol. (Inst. Pasteur) 125C: 373 (974).
22. K. Eichmann, Eur. J. Immunol. 5: 511 (1975).
23. H. Frischknecht, H. Binz and H. Wigzell, J. Exp. Med. 147: 500
    (1978).
24. R. Woodland and H. Cantor, Eur. J. Immunol. 8: 600 (1978).
25. D. Hetzelberger and K. Eichmann, Eur. J. Immunol. 8: 846 (1978).
26. K. Eichmann, I. Falk and K. Rajewsky, Eur. J. Immunol. 8: 853
    (1978).

Acquired radioresistance, 86
Actimic keratosis, 113
Actinomycin D, 125
Adenylate cyclase
     stimulation, 171
Adriamycin, 111, 112, 125
Agaricus bisporus, 7
Age effect on repair, 112
Agglutinin
  agaricus, 7
  peanut, 4, 168
  ricinus, 7
  sophora, 8
  soybean, 4
  wistaria, 7
Aggregates of lectins, 8
Ames test, 135
Anti-idiotypic antisera, 178
AP-endonuclease, 148
Aphidicolin, 38
Apurinic site, 142, 147
Apyrimidinic site, 147
Arabinosyl-cytosine (ara-C), 56
Assay for functional DNA
     repair, 98
Ataxia telagiectasia, 39, 102,
     103, 132
Autoantibodies, 165
Autologous anti-idiotypic
     antibody, 180

Basal cell carcinoma, 113-114
Basal cell naevous syndrome,
     113-114
B cell
  maturation, 163-164
  UV light sensitivity, 73, 80
Bidirectional DNA replication,
     16, 17
Binding
  affinity, 6
  cooperative, 7
  site, 8
  specificity, 4
Bleomycin, 39, 125
Bloom's syndrome, 39
BrdU, 120

cAMP
  cGMP ratio, 168
  content, 58
  levels in thymocytes, 168, 171
  second messenger, 167
Calf thymus enzymes in DNA
     repair, 146
Cap formation, 10
Capping, 9
Carrier-transport system of
     dThdr, 90
CCNU, 125
Cell surface receptor, 177
cGMP, 168